Learn Game Theory

A Primer to Strategic Thinking and Advanced Decision-Making.

By Albert Rutherford

Copyright © 2021 by Albert Rutherford. All rights reserved.

No part of this publication may be reproduced, stored in a retrieval system, or transmitted in any form or by any means, electronic, mechanical, photocopying, recording, scanning or otherwise, except as permitted under Section 107 or 108 of the 1976 United States Copyright Act, without the prior written permission of the author.

Limit of Liability/ Disclaimer of Warranty: The author makes no representations or warranties with respect to the accuracy or completeness of the contents of this work and specifically disclaims all warranties, including without limitation warranties of fitness for a particular purpose. No warranty may be created or extended by sales or promotional materials. The advice and recipes contained herein may not be suitable for everyone. This work is sold with the understanding that the author is not engaged in rendering medical, legal or other professional advice or services. If professional assistance is required, the services of a competent professional person should be sought. The author shall not be liable for damages arising herefrom. The fact that an individual, organization of website is referred to in this work as a citation and/or potential source of further information does not mean that the author endorses the information the individual, organization to website may provide or recommendations they/it may make. Further, readers should be aware that Internet websites listed in this work might have changed or disappeared between when this work was written and when it is read.

For general information on the products and services or to obtain technical support, please contact the author.

For your FREE GIFT: The Art of Asking Powerful Questions in the World of Systems visit www.albertrutherford.com.

Table of Contents

Table Of Contents

My Story With Game Theory

Game Theory Basics

Chapter 1: The Prisoner's Dilemma

Chapter 2: The Nash Equilibrium

Chapter 3: The Mixed-Strategy Nash Equilibrium

Chapter 4: Mixed-Strategy Algorithm

Chapter 5: Pure And Mixed Nash Equilibria: Mixed

Chapter 6: Strict And Weak Dominance

Chapter 7: Curious Tales From The Land Of Game Theory

Exercise

Conclusion

Resources

Endnotes

game[i]

1. a physical or mental competition conducted according to rules with the participants in direct opposition to each other.

2. an activity engaged in for diversion or amusement.

3. a procedure or strategy for gaining an end.

theory

1. a plausible or scientifically acceptable general principle or body of principles offered to explain phenomena.[ii]

My Story with Game Theory

Game theory ended my first marriage. My ex-wife had champagne tastes, but we were on a beer budget back in the day. I had just gotten my master's degree at the university I was attending, and she just didn't get the role after another audition.

To celebrate my graduation and cheer her up, we went out for dinner with our larger friend circle. After exchanging some conversational niceties, our focus shifted onto the menu. My ex-wife realized that the restaurant served her favorite dish, Moroccan lamb with a merlot glaze, sprinkled with cherry reduction sauce and paired with oven-roasted rosemary pears with gorgonzola cheese. In short, it had every possible high-end ingredient, with a price tag that kept the restaurant in business even if they sold two of the fancy mix a night. Let's say it cost $150. (And in the '80s, that was something.)

My ex-wife had a big decision to make: go for the lamb dish or not. In practice, this was a relatively simple choice. She had to decide whether indulging in the meal was worth the hefty price tag. Her to-order-or-not-to-order dilemma has little to do with game theory thus far.

But let's not forget that she was not the only person in our group. Including the two of us, we were a party of ten, all agreeing to be gentle people and split the bill evenly. As my ex-wife and I were the ones who assembled the party, we patiently waited for everyone to make their order: American hamburger (no extra cheese); Greek salad; French fries; Italian sausage; Brazilian coffee; nothing, thank you; and so on. Everybody was considerate about what the others ordered. And then my ex-wife dropped the bomb, Moroccan lamb. Boom.

Now, from an economic point of view, her decision was smart. She could treat herself to the magnificent Moroccan meal for a tenth of its price. Analyzing her behavior from a purely strategic angle, what would you say? Did she make the right call? What do you think happened next?

Well, Newton's third law kicked in, that's what happened. For her action, there was an opposite and (almost) equal reaction.

One of our friends called the waiter back. Suddenly, everyone seemed to discover that there was another page on the menu. French fries turned into Beef bourguignon and Italian sausages into risotto con porcini e tartufo. Our humble celebratory get-together started to resemble a vendetta feast of Russian mobsters. *Where is our Dom Pérignon?*

After bleeding the expensive side of the menu out of options, I finally asked for the bill. Once we split it equally, the result was $342 per person! My ex-wife made a mistake of both an economic and social nature. But was she the only one? Being irritated by her inconsiderate order and trying to make a point, everybody ended up paying much more than if they just bit the bullet and paid $15 each for her Moroccan lamb. What would you have done? What do you think my friend group should have done? I'll let you decide.

My ex-wife decided that everybody was mean, including me who ended up paying for both our shares, and she had no interest in sticking around our group—or me—any longer.

The case of the restaurant shed light on the influence interactions of various decision-makers can have on each other.

How I Learned about Game Theory

I was a nerd ever since I was a kid. And when I say nerdy, I mean curious. I would disassemble my grandparents' precious Zenith radio just to see what was inside of it. How did it work? Why did it work that way? I asked endless questions about gravity, and buoyancy. What fascinated me the most was not the laws of nature and science affecting them but how human beings were able to discover and describe them.

And then it was chess. My first love at the age of six. It was much more exciting than the radio, which was hooked up on a circuit of metal chords producing a predictable outcome. Or gravity, which, while it is a complex physical phenomenon, is fairly predictable. If you drop something in a space affected by gravity, it will fall down. The speed of the fall and force of the impact depends on the object's size and mass, but it will fall, nevertheless.

Chess was different. The outcome of the game is unpredictable the moment two people sit down to play. Who wins depends on one decision at a time. And each future choice evolves as a function of the present one. Even as a child, I was fascinated by the strategic thinking involved in chess. And again, I was mostly drawn to the question, how did human beings invent such a complex game? What was their motivation when creating chess? How do the laws of chess play out on a larger scale?

Seeking answers to these questions over the next decades was one of the best decisions of my life. On my journey, I came across Sun Tzu, Jon von Neumann, Oscar Morgenstern, John Nash, and others. Even though I couldn't grasp every concept right off the bat, I was thrilled. A deep passion fueled my willpower to learn more about strategic thinking. I was devouring the written explanations of the sometimes jargon-filled pages. But as much as I wanted to advance my knowledge, I couldn't understand the complex mathematical formulas. Not even the simple ones.

I was diagnosed with dyscalculia. In the mid-'70s, to even get this diagnosis was a miracle. This learning disability was barely recognized in those times. I was extremely lucky to have a progressive doctor. But as a child hungry for knowledge—mathematical knowledge—I felt devastated. In a fraction of a second, my dreams of becoming a great strategic thinker vanished.

When I got home, my mom embraced me tightly.

"I will never be a great mathematician like John Nash!" I cried. "Or a great general like Sun Tzu!"

My mom held me tighter. Her heart was even heavier than mine. After a few moments, she released me and picked up the worn copy of *The Art of War* and she read me Sun Tzu's words: "If you know the enemy and know yourself, you need not fear the results of a hundred battles."

Suddenly, that deep burning passion hit me again. I wasn't going to give up on my dreams! This was an obstacle, yes. But I knew my heart. I knew my enemy: numbers. The shadow of hundreds and thousands of battles didn't scare me anymore. I decided that I wasn't going to be defeated, so I focused on strengthening my math skills, one division, one multiplication at a time. My mom was my champion, my fiercest ally. She sat by my side tirelessly, patiently, going over and over the addition table. By the time I was in the second grade, I was up to the same speed as my classmates. When I reached fourth grade, I could solve the complex algorithms in my beloved books on game theory.

My early diagnosis of dyscalculia was a blessing in disguise. If I wasn't so keen on learning about strategic thinking and calculations at such a young age, my learning deficit would have showed itself at some point in school. I would have been labeled a child with special needs, or simply dumb. Had I had a shortsighted doctor or a less caring mother, I wouldn't be writing this book for you today. I feel an immense gratitude for their help. I owe many thanks to my mentors, Sun Tzu, John Nash, Jon von Neumann, and Oscar Morgenstern.

I would like to shout out to you, who has or might know someone who has dyslexia or dyscalculia. It's not a death sentence. It's a barrier. An additional challenge we didn't ask for but we got it. By no means does this make you "dumb" or "less than." You can achieve just as much as those who don't have these obstacles. It will take you more work, yes. But it's not impossible. And the benefits of rigorous study and practice will compound in your life. You will learn discipline, master focus, and you won't be a stranger to growth under adversity. These are such valuable life skills! Stay strong, work hard, and remember, "victory usually goes to the army who has better trained officers and men."

Why I Wrote This Book

I have a lifetime invested in the study and application of game theory principles. Being able to assess decisions based on cold mathematical calculations helped me make more informed, better decisions. In some aspects of life, one should follow their heart. These areas include relationships, friendships, and what wall paint color to choose. But the majority of major life decisions require rational, deliberate, and strategic thinking. Emotional responses to these problems lead to unpredictable outcomes at best. More likely, they bring chaos into our lives. Which areas am I talking about? Finances, work-related decisions (including what career to choose), health choices, purchasing options, political activity, and so on. Even the life areas that initially rely on emotions call for rational thinking after a while. Whom to marry? Which friendships are worth pursuing? Where do you draw boundaries? This list is not exhaustive.

I learned invaluable lessons from the great strategic thinkers of the world. And I would like to pass down their knowledge to you in a more digestible and relatable way. I recommend that you read the work of every single author I mentioned above. Their work can be hard to follow unless you have a strong mind for numbers. My book is a primer, an anteroom for the heavy guns of the big guys. I tried my best to make my writing engaging, relatable, and useful. You will be able to apply the game theory principles presented in the following chapters in your everyday life even if this is the only book you read on the subject.

How This Book Will Help You

I learned the ideas and strategies mentioned here five decades ago and have relied on them all my life. I am not a millionaire, but I live comfortably. I have a supporting, loving, smart, compassionate, and caring wife, two children who mean the world to me, a dependable circle of good friends. In my work, I was respected; I had a good relationship with my superiors, colleagues, and students. I have a good life. My peace of mind is not disturbed by sleepless nights, agonizing about regrets, bad decisions, or missed chances. I can claim with certainty that I owe many of these blessings thanks to the strategic principles I am about to present to you.

In the chapters that follow, I will introduce you to game theory. I will share step-by-step strategies and plans for smarter decision-making that will benefit you for a lifetime. The book is not an academic research paper; rather, a reference and practice guide. You will come across plenty of practical advice as I describe how to apply strategic thinking in your daily choices in an easy-to-understand way. How do you find consensus in choosing a date location with your significant other? What's the best choice to get free money? How do you maximize your gains when hunting? How do you flip coins successfully?

The fields I incorporate—mathematics, economics, politics, social sciences, philosophy, and cognitive psychology—are all important to familiarize yourself with. This book will give you a synthesis of the best practices these subjects have to offer you, a growth-oriented decision-maker. These practices have been around long before I even learned how to write. Please attribute the smart things on these pages to the geniuses who discovered, unpacked, and penned them down. If you read anything silly, that's probably my mistake.

There is no one right way to learn strategic thinking, but this book will present the best method I know. And that's game theory. My examples will come in handy to anyone searching for step-by-step strategic tools for improvement, whether your goals are work, money, relationship, or human interactions related. As long as your learning target involves human decision-making, this book will be helpful.

Game Theory Basics

Game theory is the study of strategic decision-making; a framework for using mathematical models to understand the behavior and motivation of competing, rational players.[iii] Robert Aumann, the 2005 Noble Prize winner in Economics, defined game theory as "a mathematical formalization of interactive decision-making."[iv] Let's pause here for a moment. These definitions are starting to get gaudier and gaudier.

In simple terms, game theory is the formalized description of what happens between rational decision-makers. I would like to emphasize the word rational. It carries the assumption that players want to maximize their benefits, no matter what. It is a safe assumption, right? Actually, later in this book, we will see that people are not always rational about strategic decisions.

You may think that only high-stakes poker players or a mastermind of political intrigue like Karl "The Architect" Rove would use game theory. Or a super-competitive sports coach who clandestinely video records the opposing team's practice drills or fudges a birthdate to have an older player in the wrong age division. Friends, enemies, small-scale or large-scale influencers, politicians, gamers, and average people like me all benefit from understanding game theory. But here comes a tricky part. As a player, it's difficult to assess what would be beneficial to the other players. Often, you don't even know what you want to achieve.

The classic assumption here is that the reward we're shooting for is money or victory. But this is a reductionist view. There are plenty of other satisfying benefits a player can get from playing "the game": pride, positive feedback, influence, more social media followers, etc. Similarly, there are negative consequences that players risk when engaging in "the game," such as lost time, failure, lowered self-esteem, or fines.

When other people are involved in playing the game, it seems like an impossible task to predict their next move. This is where the assumption of rational decision-making comes in handy; our best bet is to assume that everybody else is just as intelligent, selfish, and victory oriented as we are. Don't expect they will let you enjoy your Moroccan lamb dish while they rock-paper-scissors for the last piece of French fries, happily chipping in to pay for your monster meal.

Game theory was coined in 1940 by John von Neumann and Oscar Morgenstern. Its importance has grown ever since. A good metric supporting this statement is that since the 1970s, twelve leading economists won the Nobel Prize in Economic Sciences for their contributions in game theory.[v] If you're a budding economist with great ambitions, there is an aspiration for you.

Game theory is present in multiple fields where logic is the governing factor:

business, finance, economics, politics, sociology, and psychology. It provides strategic thinking patterns, moves, and explanations to help you develop exceptional decision-making skills. The principles presented in this book—the famous and the lesser-known—will help you make educated guesses and decide the best strategy in competitive situations.

You would be surprised how many everyday life decisions require the process of game theory. Without realizing it, you may be using the Nash equilibrium when deciding who washes the dishes. Or playing a zero-sum game when changing traffic lanes. Most of your decisions incorporate some aspect of game theory. You want to maximize your benefits in any situation, don't you? You wish to create a scenario where your costs are at a minimum, right? That's game theory for you.

Before you get too comfortable with the notion that you are a rational decision-maker and all you have to do is master the principles presented here to become the next Jeff Bezos, I have some bad news, which can be good news. There is growing evidence in the psychological community that human beings are emotion-driven, *irrational* decision-makers.[vi] That's problematic, isn't it? Now to the good news: If that's the case, it is better to be aware of it and count on not making rational gut decisions. But knowing how to make rational decisions (this is where game theory comes handy) and being aware of your cognitive biases can actually produce superior results. Together, these two skills (self-examination for irrationality and high-level strategic thinking methods) will turn you into a fierce player. It doesn't matter if you are a political or business negotiator, a street haggler, product pricing specialist, auction designer, high school teacher, military strategist, or a passionate board gamer; you will excel at decision-making.

Before I elaborate more on skill two, game theory, I want to briefly talk about skill one, irrationality detection.

Predictably Irrational Creatures

The idea of maximum benefits at a minimum cost is not new. In fact, it is as old as human nature. It's coded in our genetic makeup to seek pleasure and avoid pain. To understand the roots of game theory, we need to go back to the 18th century, the birthdate of classical economics. In 1776, Adam Smith published a groundbreaking book, *An Inquiry into the Nature and Causes of the Wealth of Nations*, discussing concepts such as rational decision-makers and rational choice theory.

"Rational choice theory assumes that individuals are rational actors using rational information to try to actively maximize their advantage in any situation and therefore consistently trying to minimize their losses."[vii]

In the next centuries, classical economics and rational choice theory dominated the field of economics. These concepts also shaped how people thought about themselves as decision-makers. They assumed that however they were thinking, whatever decisions they made, it was a calculated, rational choice.

The 20th century brought a change. The field of psychology evolved rapidly and new discoveries brought up some questions regarding human rationale. Increasing data proved that humans don't always act in their best interest. We are emotional beings who get easily distracted. Our subjective needs and wants aren't always in alignment with what's objectively beneficial. We are irrational decision-makers. These ideas were introduced by great contemporary thinkers such as Herbert A. Simon and Daniel Kahneman. The field of study incorporating Simon's bounded rationality theory (individuals' rationality is limited by the tractability of the decision problem) and

Kahneman's prospect theory (individuals assess their loss and gain perspectives in an asymmetric manner) is called behavioral economics. Game theory and behavioral economics are closely intertwined. Behavioral game theory aims to analyze people's irrational decisions through experiments.[viii]

This book is a classical game theory primer. The cognitive strategies here will aid both the rational and irrational mind. For the former, it is a great asset for smarter thinking; for the latter, it is an invaluable self-checking tool.

The Father of Game Theory

We can't talk about game theory without mentioning John von Neumann. He was a Hungarian-American polymath who contributed generously to the fields of mathematics, physics, statistics, computing, and economics (focused on game theory). During WWII, he worked side by side with Edward Teller and Stanislaw Ulam on the Manhattan Project, aiding in the discovery of the hydrogen bomb.[ix]

Von Neumann founded the field of game theory. He stated that in zero-sum games, where players are aware of every move taken so far, there is a strategy for each player to minimize their losses and maximize their gains. Player A has to analyze every strategy available and consider each possible step of Player B. Player A then will choose the strategy that minimizes their maximum loss.[x]

Von Neumann's model continued to expand to include non-zero-sum games, or games with incomplete information, and games with more than two players. This led to the publication of *Theory of Games and Economic Behavior* in 1944 with Oskar Morgenstern. Following the publication of this book, game theory entered a period of rapid and extensive development. It was applied and regularly used in topics such as evolution, social science, economics, and computer science.

Let's take a quick look at what the most popular types of games in game theory are.

- <u>Cooperative and Non-cooperative Games</u>—Some games encourage you to form alliances. These partnerships can sometimes be the most unlikely ones, as we could see throughout the show *Game of Thrones*. What matters is that you and your allies commit to having some obligations to one another.[xi] Game theory's main primary purpose is to predict which players will form an alliance, what joint action they will take, and what gains they can expect. In the real world, coalitions, merges, oligopolies, and trusts can be illustrative examples. In this book, we will focus on non-cooperative games. These games aim to predict individual players' actions and payoffs. In non-cooperative games, participants either can't form alliances due to the nature of the game or the agreements they make are based on credible threats which they need to self-enforce.[xii]

- <u>Symmetric/Asymmetric Games</u>—In symmetric games, the payoffs of a strategy depend only on the other strategies employed, not on who is playing them.[xiii] In other words, if players can be replaced without changing the gains or losses in the strategy, the game is symmetric. The prisoner's dilemma and stag hunt are two classic examples of this. We will learn about both later.

    Asymmetric games have different strategies for each

player. The ultimatum game is one such game. This was a popular economic experiment proposed by the Hungarian-American John Harsányi in 1961.[xiv] In this game, one player, the proposer, gets x amount of money. Then, the proposer needs to split it with another player, called the responder. Once the proposer chooses the amount and shares the news with the responder, they may accept it or reject it. If the responder accepts, the money is divided per the proposer's wishes. If the responder rejects, both of them get nothing. Both players know in advance what happens if the responder accepts or declines the offer.[xv]

Some games can have the same strategy for both players and still be asymmetric.

- <u>Zero-Sum Games/Non-Zero-Sum Games</u>—We talk about zero-sum games when there is a constant and unchanging amount of a resource that player decisions won't influence. Depending on the strategies used, each player gets more or less of the available resource. In other words, the total benefit is shared in various ways among all players and always adds to zero. One player gains at the equal loss of other players.[xvi]

Imagine a large pizza, which is 14 inches of cheesy goodness. Depending on who slices it, one diner might get that weirdly small slice that can be eaten in two bites while another diner gets a slice as big as their head. The total amount of pizza is still 14 inches, but the second diner got more of their fair share. Matching pennies is a classic zero-sum game in game theory, and we will learn about it later. Chess and go are the famous zero-sum games in the board game world.

Game theorists love studying non-zero-sum games like the prisoner's dilemma. Here, the outcome has net results bigger or smaller than zero. In a non-zero-sum game, one player's gain does not necessarily equal the loss of other players.

- <u>Simultaneous/Sequential Games</u>—In simultaneous games, players either make their move at the same time or, in case they don't move at the same time, they are blind to the other players' moves. The card game War is an excellent example of simultaneous games as it has players simultaneously showing their top card, with their highest card winning. The goal of the game is for one player to win all fifty-two cards.

In sequential (or dynamic) games, you know about previous actions of other players. Chess is a sequential game. The information a player has is often not perfect or complete. It may, in fact, be very little knowledge. Think about the game Battleship.

Another difference between simultaneous and sequential games is that simultaneous games often use normal form, while sequential games use extensive form.

- <u>Perfect/Imperfect Information</u>—In a game with perfect information, all players know all previous moves of other players.

When you play with imperfect information, on the other hand, some information about the moves is missing. Perfect information games include tic-tac-toe, checkers, and go. Imperfect information games are poker and bridge.[xvii]

- <u>Complete/Incomplete Information</u>—In a game with complete information, the players know the structure of the game. What does this mean? The players know the order in which they are supposed to move. They know every possible move in each position and the payoffs for every outcome. Games in the real world are usually not like this; they don't have complete information. In game theory modelling, we assume complete information as games with incomplete information are harder to analyze.[xviii]

Game Representation

The games you find in game theory are well-defined mathematical tools. We need three things to define these games:

- "the players of the game,
- the information and actions available to each player at each decision point,
- and the payoffs for each outcome."[xix]

Game theorists then use the three attributes above, plus their choice of solution plan for the game, to conclude what equilibrium strategies each player should use. In their deduction, they make sure that no player can get an advantage by unilaterally diverging from the main strategy. The equilibrium strategies define the game's equilibrium point, which is a stable condition. The probability of one or more outcomes happening is known.

In the previous section, I mentioned that simultaneous games use normal form and sequential games use extensive form. Let's take a closer look at what these terms mean.

Normal Form

Normal-form games are illustrated as tables or matrices. The number of rows equals the number of Player 1's strategies. The number of columns is the number of Player 2's strategies. The matrix, therefore, shows all of the potential outcomes based on every possible strategy a player could use.[xx]

Jim ↓ / Tim →	**Left**	**Right**
Up	4; 2	-2; -2
Down	0; 0	2; 4

Table 1 presents a two-player game. The actions of Jim are presented horizontally, and of Tim vertically. Each player has two strategies. This is illustrated by the number of rows and columns. The payoffs are the numbers in each box. The first number is the payoff of Jim. The second number is the payoff of Tim. If Jim chooses the strategy to play Up and Tim plays Left, then Jim receives a payoff of 4, and Tim gets 2. Why did the players choose this strategy instead of another one? We will learn about it in detail in the following chapter.

For now, let's just keep in mind that when a game is illustrated in normal form, we expect each player to act simultaneously. Or, at the very least, without knowing the moves of the other. If players have some information about the moves of other players, we are usually talking about a sequential game which uses the extensive form.

Extensive Form

Extensive form describes games using game trees. It's a diagram that shows that decisions are made at different points in time. The nodes show the point when a choice was made. The payoffs are shown at the bottom of each branch. The game trees help players to predict all decisions and counter-decisions in any game. We need to use backward induction to solve these kinds of games. What does this mean? We work our way bottom-up on the game tree, trying to identify what a rational Player A would do at each node of the branches, what Player B would do given that Player A's last move was rational, and so on until we reach the first node of the tree.[xxi]

An extensive form game

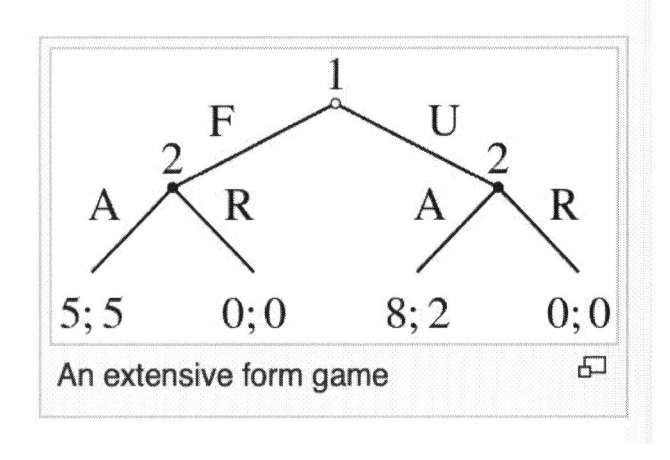

Picture 1: Extensive form game.[xxii]

The game in Picture 1 has two players. This game has sequential decision-making and perfect information. Meaning, one player goes after the other and each player knows the move of the previous player. Player 1 can go to either F or U (fair or unfair). Depending on the choice of Player 1, Player 2 can choose between A or R. Once

Player 2 is done, the game is also finished. Each of them gets their payoff. If Player 1 chose U and then Player 2 chose A, Player 1 receives a payoff of "eight" (eight thousand dollars, eight donuts, eight days of paid vacation, etc.) and Player 2 gets "two" of the payoff.

We won't be exploring extensive-form games in this book as they are a little more advanced. This book is a primer and will focus on simultaneous games with normal form.

Game Theory in the Real World

In the mid-1800s, Antoine Augustin Cournot developed an economic theory inspired by monopolies, oligopolies, and duopolies (a special type of oligopoly with only two competitors). We still rely on the principle of Cournot's competition today. The basic premise of this model was that, when two or more companies sell a product with no differentiation, companies cannot cooperate. Each company's output decision affects the market and changes prices. If Company A floods the market with a product, that product will see a decline in prices as there is a saturation of availability. If Company A's competitors don't reduce their prices, they will be out of business. However, if they are forced to reduce prices against their will, that can be the beginning of a market war where output increases and prices decrease, and in the end, no one wins. On the other hand, if there are too few products available, the prices can be drastically inflated.

Take the diamond market as an example. Thank to excellent advertising, "diamonds are forever," the general public has been led to believe diamonds are rare and special gems, the symbol of eternal love. This implanted belief legitimized the extravagant price tag. However, global diamond corporation, De Beers, has controlled the diamond market since the 1980s. They have kept pricing and quality of the stones with a ruthless grip. De Beers has done this by flooding the market with below-cost stones in retaliation for sightholders attempting to start up their own diamond companies when they believed De Beers's rules of business were overly strict and bad business.

Because De Beers controlled about 85% of the diamond market at one time (it is now less than 30%), independent-minded individuals were beholden to the company, which in the case of De Beers, meant paying the set price without any negotiation, making any company available to De Beers for audit, and accepting diamond purchases as lots whether the stones in the lots were of good quality or not. These were just a few of the practices regularly experienced by sightholders.[xxiii]

On the other hand, consider the differences between chicken nuggets purchased at a fast food restaurant. The top three options in the US are probably McDonald's, Burger King, and Wendy's. While these chicken nuggets aren't priced exactly the same, they cost more or less the same. These fast food retailers have to keep their prices relatively steady because there are other chicken nugget vendors like Chick-fil-A or KFC that eat into the market.

The earliest forms of game theory gave way to Darwinism and were used to explain evolution and animal behavior in the biological sciences. In soft sciences such as psychology or sociology, game theory is used to determine and predict both behavior and normative behavior. By listing out particular actions, it helps individuals determine what actions are ethical and acceptable in particular situations and those that are not. Game theory has also been expanded into subcategories such as chemical game theory, where chemical reactions are predicted as if the players had actually carried through with those chemical reactions.

Game theory has a place in any realm where logic and strategy are used. Businesses and corporations use it to determine how to price their products or spend their advertising dollars. Politicians and elected officials may use game theory to determine how to approach their campaign and how to win against their competitors.

The aim of this book is to explain and help you understand game theory and how you can apply the science of strategy into your life in order to benefit from and make strategic predictions. By making strategic decisions economically and socially, you will be able to confidently make predications and anticipate counteractions of other "players" when you engage in the variety of games life throws your way. The math won't get too complicated, and you know have the basics of the history and foundations of game theory to get you started. Without further ado, let's go to jail.

Chapter 1: The Prisoner's Dilemma

The prisoner's dilemma is one of the most famous examples of game theory to ever be created. Melvin Dresher and Merrill Flood conducted the first experiment based on the concept of this dilemma for the RAND Corporation in the 1950s.[xxiv] Let's take a look how the traditional—which only means the most popular—prisoner's dilemma game looks like.

Standard Prisoner's Dilemma[xxv]

Two criminals, Jim and Tim, have been picked up by the police to be interrogated on suspicion of committing a major crime, let's say, manufacturing methamphetamine. The police has no evidence, but they are certain of Jim and Tim's guilt. The two criminals are led into separate interrogation rooms by police officers who offer each one a deal. They are told that if both of them stay silent, they will have to serve one year each on a lighter charge, such as marijuana possession. This is not the best deal, is it? But then the chief of police enters the room, and as a part of a well-rehearsed police strategy, gives them each a better deal. If Jim rats out Tim, he gets immunity for his testimony. But Tim will serve ten years in prison, now for a provable crime. Tim gets the same offer. If both talk, they will both serve eight years having confessed to their crimes. Should either criminal talk or remain silent? Jim and Tim have to make their decision without knowing the decision of the other one.

Jim ↓/Tim →	**Stays Silent**	**Betrays**
Stays Silent	-1; -1	-10; 0
Betrays	0; -10	-8; -8

The table above shows the rules of the game. The numbers are the number of years spent in prison. Mathematicians call these tables a game matrix. So far, we don't have much to say about this matrix other than, "Why can't we just call it a table?" This story gets exciting once we start solving it.

At first glance, which solution seems the best? Of course, the one where they both stay silent. They serve a year at good taxpayers' expense, and after that, go to live on a Caribbean island with the meth money they already laundered through a nail salon in Texas. But, as the old saying goes, there is no loyalty among thieves. And Jim certainly wouldn't swear on Tim's trustworthiness. He eats his steaks well-done after all …

"I don't know what Tim will say, but I know that I have only two choices: to speak or not to speak. Very Shakespearean. If Tim and I both stay silent, we both serve one year. But … but if Tim opts for the Fifth[1] and I talk, I walk! I should talk," ponders Jim. *"However, if Tim is a rat and I don't say a thing, I will rot in jail for ten years! If I talk, at least I will only serve eight."* And then slowly it sinks in to Jim: *"Whatever choice*

---

[1] "In criminal cases, the Fifth Amendment guarantees the right to a grand jury, forbids "double jeopardy," and protects against self-incrimination." Cornell Law School

Tim makes, I am better off if I talk either way!"

Jim's monologue sums up the main message of the traditional prisoner's dilemma. Strategically speaking, Jim is better off talking. If Tim stays silent, Jim gets no punishment. If Tim talks, well, that's still eight years instead of ten.

Remember the concept of symmetric games we talked about in the introductory chapter? This is a prime example of it. Both players are equal. Tim had the very same deduction monologue with himself as Jim. He reached the same conclusion. They made a rational, self-protecting decision, namely, they betrayed each other, yet they both ended up in jail for eight years. A cruel game, isn't it?

Besides being a symmetric game, the prisoner's dilemma is also a game with a strictly dominant strategy. What does this mean? Let's go through Jim's choices again assuming Tim stays quiet:

Jim ↓ / Tim →	**Stays Silent**
Stays Silent	-1; -1
Betrays	0; -10

If Tim stays silent and Jim also stays silent, the outcome for Jim is 1. If Tim is silent but Jim talks, the outcome for Jim is 0. In other words, the strategy where Jim speaks strictly dominates the one where Jim stays silent.

Let's see Jim's choices if Tim betrays:

Jim ↓ / Tim →	**Betrays**
Stays Silent	-10; 0
Betrays	-8; -8

If Tim speaks and Jim stays silent, that's a 10 for Jim. Not good. If Tim and Jim both speak, Jim gets 8. Therefore, the strategy where Jim talks, again, strictly dominates the one where Jim remains a good friend. In other words, in both cases, keeping silent is a strictly *dominated* strategy. No rational player wants to ever play a strictly dominated strategy.

> If we went through the logical deduction of this game from Tim's perspective, we would get the same results: The strategy of betrayal

strictly dominates the strategy of staying silent.

Where did Jim and Tim go wrong? The sensible and logical decision is for both to confess. Yet, neither of them got into a better position. In fact, both added seven years to what could have been only one.

If you showed the game matrix of the prisoner's dilemma to people who don't know the logical deduction, they may be perplexed why Jim and Tim didn't choose the -1; -1 option. If both kept quiet, they both would have been better off. And that is true. Serving one year only is better both individually and collectively than confessing. The -1; -1 outcome is not sensible because, in that case, each of them is better off individually if they confess. So,

-1; -1 doesn't offer a stable, reliable solution. Because even if Jim and Tim agreed to not rat each other out beforehand, they could always flip as they go to improve their individual outcomes.

The outcome of the prisoner's dilemma game is inefficient. But we also assumed that both Jim and Tim made their decisions based on the sole wish to diminish their prison sentence. If other things come under consideration, like a big meth boss killing anyone who spills the beans to the police, they may have opted for different choices.

Economic Prisoner's Dilemma[xxvi]

Let's examine the prisoner's dilemma playing out in business. Both Nike and

Adidas are top manufacturers of running shoes, and there is an implicit agreement between both companies for an average price for low-, mid-range, and top-tier shoes. However, suppose Nike is considering cutting prices, which is tantamount to defecting on the implied agreement between the athletic shoe companies. This means Adidas would also have to cut their prices to remain competitive or risk losing its customers to Nike. If this happens, both companies can lose out bigtime as profits plummet. If Adidas keeps its prices up at the original level in the wake of a Nike defection, Adidas has remained true to the intent of the original agreement. Thus, because Nike has violated the spirit of the agreement, they may gain a higher percentage of the market share. Let's assume the following …

- If both companies maintain their higher prices, their profits will increase by $250 million.
- If one company defects and the other cooperates by not dropping prices, the company that raises prices will increase profits to $500 million while the cooperating company experiences no gain.
- If both companies drop their prices (defect), then both companies experience a profit increase of $125 million.

We can view the potential outcomes and payoffs for this scenario in a matrix.

Adidas↓/Nike →	**Cooperate**	**Defect**
Cooperate	$250k; $250k	0; $500k
Defect	$500k; 0	$125k; $125k

This is just one example of how the decisions of players in a game can influence and predict the actions of other players. Even advertising can have profound impact on a company's bottom line and impact net profit. For example, companies like Doritos (owned by Frito Lay) and Budweiser (owned by Anheuser-Busch) are known for their annual Superbowl commercials. Many people watch the Superbowl as much for these expensive and elaborate commercials as they do for the enjoyment of the game. It's simply understood that there will be commercials from particular companies and Budweiser is famous for including

Clydesdales, frogs, and bikini-clad women in their ads.

However, in 2021, Budweiser threw everyone a curveball and announced the week just prior to the championship game they had not purchased ad space for the Superbowl this year. They officially defected. All other beer companies will see Budweiser profits increase because the King of Beers has violated the implicit agreement that they will spend major advertising dollars for this event. Unless they dump a significant amount of money elsewhere, Budweiser may have an increase in profits, global pandemics notwithstanding.

The Prisoner's Dilemma in the Real World

Let's talk about asking for that long-overdue raise at your job. In our game matrix, we will measure the level of satisfaction of you and your boss.

You↓/ Boss→	**Cooperate**	**Defect**
Cooperate	6; 6	0; 10
Defect	10; 0	3; 3

What does this matrix tell us? It is generally not advised to take the first offer an employer gives and leave room for negotiation. You know you're worth more. Cooperation might seem tempting. You will get a bit of a salary raise and your boss will also be somewhat satisfied. They got you off their back for another two years. You could try to decline your boss's offer, asking for more. You may get your raise but that would leave your boss pretty pissed, looking for your first misstep to punish you. Conversely, if your boss deflected your request for a raise, they might keep the money but you would be deeply unsatisfied. Your hardworking spirit would decline. If both of you negotiated, you may end up in a somewhat satisfying situation. But the extensive bargaining and finger-pointing of the process would deplete both of your goodwill.

The Prisoner's Dilemma in Deadlock[xxvii]

The biggest issue with strict dominance games like the prisoner's dilemma is that their outcome is often not optimal. Both players must actively hurt the other in order to "win." In doing so, they both end up with a worse outcome. Remember Jim and Tim betraying each other in order to strike a better deal for themselves, but they will serve eight years in prison instead of one?

A deadlock follows a similar logic as the prisoner's dilemma where both players hope for the other to cooperate but neither actually does. Yet the outcome is Pareto optimal.

"Pareto optimality is a measure of efficiency. An outcome of a game is Pareto optimal if there is no other outcome that makes every player at least as well off and at least one player strictly better off. That is, a Pareto Optimal outcome cannot be improved upon without hurting at least one player."[xxviii]

Adidas↓/Nike→	**Cooperate**	**Defect**
Cooperate	$15m; $15m	$0; $30m
Defect	$30m; $0	$20m; $20m

Let's go back to our example of Nike and Adidas. Suppose they formed a duopoly of keeping prices stable, following which they would gain fifteen million dollars each. If Nike disrespects the agreement and drops its prices to pull in a larger percentage of the market share, their profit margin would grow by thirty million dollars. If both would stab each other in the back, however, and start selling cheaper shoes, they would tap into a much larger market and both would end up better, cashing in twenty million each. Giving up on their agreement, thus, works out for both of them.

How to Eliminate *Strictly Dominated* Strategies[xxix]

Let's recall the solution to the prisoner's dilemma.

Jim ↓/Tim →	**Stays Silent**	**Betrays**
Stays Silent	-1; -1	-10; 0
Betrays	0; -10	-8; -8

We concluded that the strictly dominant strategy was where both betray each other. Staying silent, on the other hand, was a *strictly dominated* strategy. Why? Because spilling the beans always produced a better outcome to the individual.

There are, however, some games where one strategy is not always better for each player. It may happen that one player changes their strategy based on what the other player is doing. In those cases, we need to apply a different analytical approach. This is what we call the iterated elimination of strictly dominated strategies.

I know, the name sounds scary. But the logic behind this method is simple. Let's look at an example:

Player 1↓ / Player 2→	**Left**	**Center**	**Right**
Up	12; 2	0; 3	6; 2
Middle	3; 0	2; 2	5; 1
Down	-2; 8	1; 7	7; -2

Don't worry about the table's complexity. It is not as hard to decipher as it may look. Let's define what we see. We have an extra row and column compared to the prisoner's dilemma matrix. This only means that Player 1 and Player 2 have three strategies in this game. Player 1 has up-middle-down and Player 2 has left-center-right.

Let's go through each strategy one by one. If Player 2 decided to play left, what would be the sensible choice for Player 1? To go up. 12 is more than 3 or -2.

Player 1↓/Player 2→	**Left**
Up	12; 2
Middle	3; 0
Down	-2; 8

If Player 2 chose "center" as strategy, Player 1's best bet is to play "middle." Why? Because 2 is bigger than 0 or 1.

Player 1↓/Player 2→	**Center**
Up	0; 3
Middle	2; 2
Down	1; 7

If Player 2 went right, Player 1 should play down. 7 is bigger than 5 or 6.

Player1↓/Player2 →	**Right**
Up	6; 2
Middle	5; 1
Down	7; -2

As you can see, based on Player 2's moves, Player 1 has different optimal strategies. How can we solve this game knowing all this? Instead of doing the same analysis from Player 1's perspective, let's take a closer look at Player 2's strategies. Would Player 2 ever choose right?

Player 1 ↓ / Player 2 →	**Center**	**Right**
Up	0; **3**	6; **2**
Middle	2; **2**	5; **1**
Down	1; **7**	7; **-2**

Look at the bolded and highlighted numbers: Player 2's strategies. We can see that regardless of Player 1's moves, Player 2 is always better off choosing "center;" 3 is bigger than 2, 2 is bigger than 1, and 7 is bigger than -2. "Center" strictly dominates "right." It makes sense for Player 2 to ditch "right," right? After the adjustment, the game would look like this from Player 2's perspective:

85

Player 1 ↓ / Player 2 →	**Left**	**Center**
Up	12; 2	0; 3
Middle	3; 0	2; 2
Down	-2; 8	1; 7

Player 1 is very astute. He assumes that Player 2 will not play right. How should Player 1 respond? If Player 2 never plays right, should Player 1 ever play down?

Player 1 ↓ / Player 2 →	**Left**	**Center**
Up	**<u>12</u>**; 2	**<u>0</u>**; 3
Middle	**<u>3</u>**; 0	**<u>2</u>**; 2
Down	**<u>-2</u>**; 8	**<u>1</u>**; 7

If you look at Player 1's options, you'll see that playing "down" is a *strictly dominated* strategy. In other words, it pays off better for Player 1 to play "middle" each time. Why? Because 3 is bigger than -2 and 2 is bigger than 1.

Let's update our game matrix based on the new information.

Player 1 ↓ / Player 2 →	**Left**	**Center**
Up	12; 2	0; 3
Middle	3; 0	2; 2

Player 2 is very intelligent. They infer that Player 1 would not play down. They look at what their next best step is.

Player 1 ↓ / Player 2 →	**Left**	**Center**
Up	12; **2**	0; **3**

| **Middle** | 3; **0** | 2; **2** |

Looking at the bolded, underlined numbers we can conclude that Player 2 is better off choosing "center" regardless if Player 1 goes "middle" or "up." 3 is bigger than 2. And 2 is bigger than 0. Moving "left" is not a good strategy for Player 2. We know she will play center.

Player 1 ↓ / Player 2 →	**Center**
Up	0; 3
Middle	2; 2

Let's recap what happened here so far:

- Player 1 inferred that Player 2 won't play right.

- Then Player 2 inferred that Player 1 won't play down.
- Then Player 1 inferred that Player 2 won't play left.

Player 1 is left with a simple decision. Middle or up? As we see, 2—moving to "middle"—is greater than 0—moving "up." Player 1 therefore will choose "middle."

We got our solution to this game:

Player 1 ↓ / Player 2 →	**Center**
Middle	2; 2

Each player will get 2. They inferred a lot about the other's strategy. This inference allowed us to discard strategy "right," then "down," then "left," then "up."

This is what iterated elimination of strictly dominated strategies means. First, "right" was strictly dominated, then "down," then "left," then "up."

When you encounter strictly dominated strategies, eliminate them. It doesn't matter in what order you eliminate them. In our game, we could have started from Player 1's point of view. We could have eliminated "down" first. Iterated elimination of strictly dominated strategies is interesting and helpful.

But unfortunately, most games don't have dominated strategies like the one we just played.

Now, you might be asking, "How can I solve those games?"

Thank you for asking, we're about to figure it out.

Chapter 2: The Nash Equilibrium

Let me introduce a famous game theory figure. His name is Nash. John Nash. You may have heard of him, as he's the main character of the 2001 film *A Beautiful Mind* starring Russell Crowe. While the film is a fictional retelling, it can be an entertaining introduction to one of the forefathers of game theory. John Nash was a brilliant man, idolizing John von Neumann, and earning his Ph.D. in 1950 from Princeton University with a 28-page dissertation on non-cooperatives games. Twenty-eight pages usually barely begins to scratch the surface of most dissertations, but Nash's contained the inner workings of the Nash equilibrium, which is a key part of non-cooperative game solutions. Nash had accomplished this all by the tender age of 21.[xxx]

At Princeton, Nash fit into the mathematics community, and he even went so far as to try to convince Albert Einstein gravity could cause friction on photons. A physicist, he was not. By the late 1960s, Nash had suffered a series of personal setbacks, including fathering an illegitimate child, being arrested for indecent exposure, and the death of his father. While it is unknown what exactly led to Nash's breakdown, he was hospitalized in 1961 and diagnosed with schizophrenia or with paranoid schizophrenia. This diagnosis led Nash to have bouts of time in and out of psychiatric hospitals over the next thirty years, suffering through insulin shock therapy and antipsychotic medication that impacted his ability to solve complex math problems. Nash's mental illness had a profound effect on his career, as he once turned down a chair position at the University of Chicago due his impending position as the Emperor of Antarctica.[xxxi,xxxii]

However, by the mid-1990s, Nash seemed to be doing better, stating he began engaging in a process of rejecting the "delusionally influenced lines of thinking." Nash also began to receive recognition for the work he had done back in the 1940s and 1950s, earning a Nobel Prize in 1994 among many other awards. Nash continued to work on game theory, specifically looking at partial agency. Tragically, John Nash and his wife, Alicia, died in a car accident on May 23, 2015 as they returned from a trip overseas where Nash had been recognized for his work and contributions in mathematics.[xxxiii]

What is the Nash Equilibrium?[xxxiv]

The Nash equilibrium is a set of strategies for each player, where no player has an incentive to change their strategy given what other players are doing. Players have control over their own choices.

Jim ↓/Tim →	**Stays Silent**	**Betrays**
Stays Silent	-1; -1	-10; 0
Betrays	0; -10	-8; -8

In the prisoner's dilemma, the outcome was a Nash equilibrium. Jim and Tim both confessed and did eight years of jail time. Tim chose to betray Jim, and it was right for Jim to do the same thing. The (-8; -8) strategy is a Nash equilibrium because if either player decided to keep silent, they would have gotten ten years instead. Thus, neither of them was incentivized to change strategies. The goal of the game is not winning but to make a decision that neither player will regret. If Jim kept silent and Tim spoke, Jim would have regretted his choice as he had to go to prison for ten years instead of eight, and vice versa. The stay silent/betray strategy pair, therefore, is not a Nash equilibrium.

In other words, the Nash equilibrium acts like a rule that no player wants to break—even if there were police around to enforce it.

The Nash equilibrium is used to analyze strategies between multiple decision-makers. When Player 1 predicts decisions, they must do so based on the decisions of others and their own. One can't do this in isolation. Player 1 must ask what the others would do considering what they expect the others to do. After getting to his own conclusion, Jim asked himself what Tim would do. Assessing Tim's side, Jim decided that he expects Tim to talk.

Nash Equilibrium in Real Life

The Nash equilibrium has been used in hostile military situations such as in the arms races, and other conflicts (the prisoner's dilemma). It can be used to show to what extent two players with differing preferences can cooperate (the battle of the sexes), or whether they will be willing to take a bigger risk to achieve cooperation (stag hunt). It can also be used to organize auctions, manage resource markets, and even penalty kicks in soccer.

Game theory is present in so many fields. You do not have to be a high-end politician, economist, or strategist to understand and apply its most influential teachings in your life. In fact, many people in the aforementioned fields have little knowledge of what I'm presenting in this book, and often they don't achieve the desired results.

The Nash equilibrium, in simple terms, is the natural human inclination to do what is best for oneself, taking as a given what others are doing. In game theory terms, a rational player is "best-responding" to other players' strategies in an individualistic and self-centered way.

For example, if there are two coffee shops near you, your best strategy buying coffee in the morning would be to go to the cheaper coffee shop. And similarly, the best strategy for the owner of the coffee shop would be to price their coffee somewhat below their competitor. That way, you would buy from them if your strategy was to buy the cheapest coffee nearby.

We understand the logic. But sometimes this logic doesn't bring us the optimal results. Think about the prisoner's dilemma. In the aforementioned movie, *A Beautiful Mind*, Professor Nash, played by

Russell Crowe, said: "Adam Smith was wrong!"[xxxv]

What did he mean? To understand it, we need to familiarize with Adam Smith's classical economic theory a bit. The famous Scottish economist declared that individuals, by following their own interest, were maximizing the collective well-being of society as a whole. He elaborated on this idea through the allegory of the invisible hand.[xxxvi] John Nash brilliantly contradicted this century-old idea with his theory. The Nash equilibrium sheds light on how acting rationally from an individual's perspective can produce collectively suboptimal (or even terrible) payoffs.

The prisoner's dilemma illustrated this well. Jim and Tim, due to the nature of the game, chose to betray each other. Betrayal was the ideal individual best response. Yet, collectively, they both ended up in prison for eight years. Had they kept silent, they would have served only one year.

The reason why this game is so well-known is that it can be utilized to explain many real-life phenomena where following one's self-interest leads to adverse collective outcome.

Think about top athletes and performance-enhancing drugs. If no one used them, everybody would be better off. Let's assume the state of no athlete using drugs would be achieved. It wouldn't take long for one of them to be tempted to secretly try using something to gain competitive advantage. (This would be equivalent to choosing betrayal in the prisoner's dilemma.) The outcome would either be a severe disadvantage for other athletes or assuming all athletes think and act alike and start taking drugs—relatively equal performance on a higher, almost unnatural scale.

The Nash equilibrium's suboptimal payoff can be applied to environmental policies, as well. On a global scale, we need to decrease CO_2 emissions and reverse the trend of climate change. But on a country level, it's against our economic interest to decrease CO_2 emission. It would mean we have less industrial output. So, each country bickers about the CO_2 emission distribution while the planet is slowly consumed by the negative effects of global warming.

What do you think about road congestion? Do you think more roads will reduce traffic jams? The answer, based on historic data and the supply-demand rule, is no. Increased supply (roads) induces more demand (more cars and traffic). Before long, the new road will be flooded by new drivers.

There is a Nash equilibrium even in traffic jams. People want to drive because it's more convenient and faster. But when everyone is thinking alike, more people will flood the roads, which leads to traffic jams. The increased supply of roads has increased the payoff for driving versus other methods of commuting. A more reasonable solution to decease road congestion could be governmentally regulated congestion fees and investing in public transport development.

How about doing that outstanding internship in college to have something to show off on your CV? If everyone went for it, no one would be outstanding on the job market. Just a bunch of graduates averaging with outstanding internships.

Examples surround us; we just have to open our eyes to these inefficiencies. The Nash equilibrium helps us understand why some problems in society need specific types of intervention to reach optimal payoffs. Legally binding regulations and multilateral talks are necessary to overcome the issues created by personal interest in cases such as global warming. Surely, this is easier said than done. On the scale and context of global economic competition where no player wants to go first in depriving themselves of production advantage, it's a tough nut to push through any regulation change.[xxxvii]

Pure-Strategy Nash Equilibrium

A pure-strategy Nash equilibrium is when players don't randomize between two or more strategies.[xxxviii] Both players are definitely choosing one strategy or definitely choosing another strategy. They don't flip a coin to decide which strategy to use. Their strategic choice is very deliberate. The stag hunt is a great example of the pure-strategy Nash equilibrium.

The Stag Hunt[xxxix]

In this scenario, Jim and Tim are hunters who, being fed up with the COVID-19 restrictions, decide to go out to hunt. In the forest, they can safely socially distance, after all. They have been informed by other hunters that there are two hares and one stag in the nearby forest. Hunting a stag calls for different equipment than hunting hares. They each need to decide what they want to hunt for and bring the correct equipment, as it requires them to work together since stag hunting is a challenge. The mighty stag has more meat than the two hares. But Jim and Tim must work together to catch it. If they don't work together, they can catch hares by themselves. The matrix for the stag hunt looks like this:

Jim ↓/Tim →	**Stag**	**Hare**
Stag	4; 4	0; 2
Hare	2; 0	1; 1

In this matrix, we see that one stag yields a total of eight units of meat. One hare provides only one unit. If both hunters decide to hunt the stag, it will result in each gaining four units of meat. On the opposite end, if both hunters go after the hares, each will get one unit of food.

If Jim chooses to hunt the stag alone, he is unsuccessful while Tim snags the hares.

Jim ↓ / Tim →	**Stag**
Hare	2; 0

Jim gets two units of meat while Tim gets none. The opposite scenario plays out the same way. If Jim goes after the stag and Tim after the hare, Jim will get zero and Tim two.

Jim ↓ / Tim →	**Hare**
Stag	0; 2

These aren't Nash equilibria because in these scenarios one player has the option to switch strategies and get a better outcome. The player who is getting zero can either choose to hunt for hare and get one. The player already hunting for hare can choose to hunt for stag and get four.

In the last chapter, we were looking for strictly dominated strategies. In this game, we will not be helped by doing that. Why? Let's say Jim knows that Tim will hunt a stag. Jim in this scenario should also go for the stag, because 4 is more than 2. (See the bolded and underlined options for Jim.)

Jim ↓ / Tim →	Stag
Stag	**4**; 4
Hare	**2**; 0

But if Jim knows that Tim will hunt for hares, he will choose to hunt for hare as well—1 is more than 0.

Jim ↓ / Tim →	Hare
Stag	**0**; 2
Hare	**1**; 1

We can conclude that Jim's optimal strategy depends on what Tim is doing. This conclusion stands true in Tim's case, too. He will hunt a hare if Jim's hunting a hare and go for the stag if Jim is also hunting for the stag. Neither of these strategies is *strictly dominated*.

How can we solve this game? Based on what we learned so far, we can't give a conclusive answer. But this is where the Nash equilibrium comes into the picture.

I mentioned before that the Nash equilibrium is a set of strategies, one for each player, where no player has any incentive to change their strategy. Nash equilibria are stable. One player's actions are optimal given what the other player is doing. In other words, once one player has chosen their strategy, they have no regrets about it. They can't do better if they change their strategy retrospectively.[xl]

Let's find out how to find the Nash equilibria (there are two) in the stag hunt game.

Jim ↓/Tim →	**Stag**	**Hare**
Stag	4; 4	0; 2
Hare	2; 0	1; 1

We will do this by looking at one outcome at the time and seeing if either players can individually do better if they change their strategies. We will begin by looking at the outcome of the Stag-Stag scenario.

Stag-Stag (4; 4)

Jim ↓ / Tim →	**Stag**
Stag	4; 4

Would either Jim or Tim have a better outcome if they choose a different strategy? Jim would not want to change his strategy because if he went for hare, he would only get two units of meat instead of four. The same stands true in Tim's case. If he went for hare, he would also end up with two units of meat only. We can conclude that the Stag-Stag outcome is a Nash equilibrium. No one has any incentive to change this strategy. Except the stag.

Looking at our game matrix, it is obvious that this is the best outcome, yielding the highest reward for both Jim and Tim. But as I dropped the spoiler earlier, there are two Nash equilibria in this game. Let's find the other one.

Stag-Hare (0; 2) or Hare-Stag (2; 0)

Let's see what happens if either Jim decides to hunt for hare and Tim for stag.

Jim ↓ / Tim →	**Stag**
Hare	2; 0

Is this a Nash equilibrium? Is this a stable strategy? No, it is not. Why? Look at the individual outcomes. If Jim knows that Tim is going to hunt the stag, it makes sense for Jim to change his strategy and also go for the stag as four is better than two. Let's approach it from Tim's perspective. If Tim knows that Jim is going to hunt for hare, he will also go for hare because one is better than zero.

The opposite scenario is also true. It's important to notice that there is an individual deviation that leaves each player better off.

Hare-Hare (1; 1)

The last scenario I would like to explore is the Hare-Hare. Do either Jim or Tim have a profitable deviation from this outcome? No, they don't. If Jim decided to go for a stag instead, he would be getting zero instead of one. The same is true in Tim's case. If he went for the stag, he would also gain zero instead of one.

Jim ↓ / Tim →	**Hare**
Hare	1; 1

Collectively (meaning in both players' case), this outcome is also a stable Nash equilibrium. This is a less obvious Nash equilibrium because, in this case, both players are worse off than if they both hunted the stag. But if there was a special reason for one of the hunters to not hunt stags, this can be a good alternative strategy to keep in mind. If both Jim and Tim follow it, they fare somewhat better than if they didn't cooperate. It's an inefficient choice, but Nash equilibria are not always efficient. They are meant to be stable.

Nash Equilibrium in More Complex Games

In the game we are about to explore, Jim and Tim are two managers competing for a promotion. They have an upcoming presentation that will decide who will get promoted. Jim and Tim each have three subordinates. They can simultaneously allocate these subordinates to help with the presentation. Or they can unilaterally withdraw from the presentation competition.

The person with most subordinates allocated wins the presentation competition as it will be a more holistic, better researched one. If they have the same amount of subordinates allocated and deliver a presentation of equal quality, the CEO will reward them both more modestly but neither will receive the major promotion. If one decides to withdraw unilaterally from the presentation competition, the CEO calls the event off as he has more important things to do than listen to just one presentation. But to keep his managers' spirit up, they both receive the same modest reward they did if they tied at the competition. In other words, if one opts out from the competition, or they tie with the subordinate allocation, they get a draw.

This game is called *safety in numbers*.[xli]

Jim ↓ / Tim →	**Pass**	**One**	**Two**	**Three**
Pass	0; 0	0; 0	0; 0	0; 0
One	0; 0	0; 0	-1; 1	-1; 1
Two	0; 0	1; -1	0; 0	-1; 1
Three	0; 0	1; 1	1; 1	0; 0

In this game, both Jim and Tim have four strategies: they either assign one, two, or three subordinates to the project, or they pass.

As you can see, if either Jim or Tim passes, that results in a draw automatically.

Jim ↓ / Tim →	Pass	One	Two	Three
Pass	0; 0	0; 0	0; 0	0; 0
One	0; 0			
Two	0; 0			
Three	0; 0			

If they allocate the same number of subordinates, they also tie. See the matrix below.

Jim ↓ / Tim →	Pass	One	Two	Three
Pass	0; 0			
One		0; 0		
Two			0; 0	
Three				0; 0

When Jim assigns more employees to the task, you can see the outcome in the boxes in dark gray. When Tim has the subordinate advantage, the outcome looks like the boxes is bright gray.

Jim ↓ / Tim →	Pass	One	Two	Three
Pass	0; 0	0; 0	0; 0	0; 0
One	0; 0	0; 0	-1; 1	-1; 1
Two	0; 0	1; -1	0; 0	-1; 1
Three	0; 0	1; -1	1; -1	0; 0

In the stag hunt game, we checked each outcome and tried to find if any of the players have a profitable deviation. We wanted to see if there was a better move for each player than the one presented in each given outcome. This type of analysis can be problematic in games with more strategies, thus more outcomes. Our game here has sixteen outcomes, but some games can have dozens and hundreds. Do we really want to check them individually?

Of course not. And we don't have to. There is a concept in game theory called "best response." What does this mean? Game Theory 101 defines best response as follows:

"Given what all players are doing, a strategy is a best response if and only if a player cannot gain more utility from switching to a different strategy. A game is in Nash Equilibrium if and only if all players are playing best responses to what the other players are doing."[xlii]

To discover the best responses in a game, we need to isolate one player's strategy and find out what the other player needs to respond with.

Jim ↓ / Tim →	Pass	One	Two	Three
Pass	0; 0	**0**; 0	0; 0	0; 0
One	0; 0	**0**; 0	-1; 1	-1; 1
Two	0; 0	**1**; -1	0; 0	-1; 1
Three	0; 0	**1**; -1	1; -1	0; 0

In the matrix above, I isolated Tim's move. He is playing the strategy where he allocates one subordinate to the presentation. Next, we need to take a look at Jim's payoffs (see the numbers highlighted and underlined). Which yield the best outcome to Jim? In case Tim allocates one subordinate, Jim is better off if he allocates two or three.

In this scenario, Tim either draws or loses. So Jim's best responses are two and three. Let's mark them with a #.

Jim ↓ / Tim →	Pass	One	Two	Three
Pass	0; 0	**0**; 0	0; 0	0; 0
One	0; 0	**0**; 0	-1; 1	-1; 1
Two	0; 0	**#1**; -1	0; 0	-1; 1
Three	0; 0	**#1**; -1	1; -1	0; 0

Now we move on to another of Jim's strategies. Let's say, the strategy where he deploys two subordinates to the presentation.

Jim ↓ / Tim →	Pass	One	Two	Three
Pass	0; 0	0; 0	**0**; 0	0; 0

One	0; 0	0; 0	**-1**; 1	-1; 1
Two	0; 0	1; -1	**0**; 0	-1; 1
Three	0; 0	1; -1	**1**; -1	0; 0

If Tim allocates two employees, what's Jim's best response? If you look at the underlined and highlighted numbers of Jim, you can see that he ties if he passes or deploys two employees himself. He loses if he only sends one, and wins if he sends three. Thus, the best response for Jim is three. Let's mark it with a #.

Jim ↓ / Tim →	Pass	One	Two	Three
Pass	0; 0	0; 0	0; 0	0; 0
One	0; 0	0; 0	-1; 1	-1; 1
Two	0; 0	**#1**; -1	0; 0	-1; 1
Three	0; 0	**#1**; -1	**#1**; -1	0; 0

We need to do this process for each player's every strategy. In this book, I will do Tim's perspective and Jim's best responses. You can try to decipher Jim's perspective and Tim's best responses before reading further.

Jim ↓ / Tim →	Pass	One	Two	Three
Pass	0; 0	0; 0	0; 0	#0; 0
One	0; 0	0; 0	-1; 1	-1; 1
Two	0; 0	#1; -1	0; 0	-1; 1
Three	0; 0	#1; -1	#1; -1	#0; 0

If Tim sends three subordinates to work on the presentation, Jim's best response will be to either pass or also send three people, or he loses. The zeros are the best responses in this case because they are more than minus one.

Jim ↓ / Tim →	**Pass**	One	Two	Three
Pass	**#0**; 0	0; 0	0; 0	**#0**; 0
One	**#0**; 0	0; 0	-1; 1	-1; 1
Two	**#0**; 0	**#1**; -1	0; 0	-1; 1
Three	**#0**; 0	**#1**; -1	**#1**; -1	**#0**; 0

Lastly, if Tim passes, any of Jim's strategies are best responses. Jim is indifferent about his strategies as they all lead to the same outcome.

Now you can see all of Jim's best responses for Tim's strategies. This is the last call to find Tim's best responses to Jim's strategies on your own. In the next matrix, I will fill them in.

Jim ↓ / Tim →	Pass	One	Two	Three
Pass	#0; #0	0; #0	0; #0	#0; #0
One	#0; 0	0; #0	-1; #1	-1; #1
Two	#0; 0	#1; -1	0; #0	-1; #1
Three	#0; #0	#1; -1	#1; -1	#0; #0

Take a look at the matrix. What can you observe? You can see that there are some strategy combinations where there is only one #. And there are strategy combinations with two #s. What does this mean?

If we recall what a Nash equilibrium means, we will see that an outcome is in a Nash equilibrium only if both (or all) players are playing best responses to what others are playing. In other words, each box with two #s contains best responses for both players.

125

The game with two #s is in Nash equilibrium.

Jim ↓ / Tim →	Pass	One	Two	Three
Pass	**#0; #0**	0; #0	0; #0	**#0; #0**
One	#0; 0	0; #0	-1; #1	-1; #1
Two	#0; 0	#1; -1	0; #0	-1; #1
Three	**#0; #0**	#1; -1	#1; -1	**#0; #0**

The pass-pass, pass-three, three-pass, and three-three boxes are the ones with mutually optimized responses. All the other responses are not "the best." One player can always benefit more.

This method is more efficient than analyzing the sixteen outcomes individually. We had to check four different outcomes in each player's case. We cut our work in half.

When you are facing more complex games with more strategies or players, it is a good bet to mark the best responses instead of going through each strategy one by one.

What Are the Limitations of Nash Equilibrium?

In real life, the main limitation of the Nash equilibrium is that each player has to know their opponent's strategy. We're talking about a Nash equilibrium only if a player sticks to their current strategy if they know their opponent's strategy.

In many instances, such as in a military war or a bidding war, players rarely know the opponent's strategy. The Nash equilibrium doesn't always lead to the most optimal payoff, it just means that a player opts for the best strategy based on the information they have.

Furthermore, in multiple games played with the same opponents, the Nash equilibrium does not take into consideration past behavior, which often predicts future behavior.[xliii]

This chapter has presented the Nash equilibrium in what is known as pure strategy. In the next chapter, we will look at the Nash equilibrium in mixed strategies. Go and grab a few coins and join me in the next chapter.

Key Takeaways
- ➢ The Nash equilibrium is the most common way to find the solution to

non-cooperative games with multiple players.

➢ There is nothing to gain by changing your strategy in Nash equilibria. Cooperation tends to net the best possible outcome.

➢ Stability is a critical requirement of the Nash equilibrium.

➢ Nash equilibria can be inefficient.

➢ When you're facing a game with multiple players or strategies, mark the best responses.

Chapter 3: The Mixed-Strategy Nash Equilibrium

When you were growing up, did you have a sibling or friend you fought with relentlessly about whose turn it was to play with a toy? What about how most opposing sports teams decide who gets control of the field first? We flip a coin and the player who isn't flipping calls out heads or tails when the coin is midair. When the coin lands, whether or not the coin is face up or down decides who gets the toy or the starter kick. But what would happen if the other player brought their own coin to this coin-flipping party, much like the villain Two-Face from *The Dark Knight*?

Assuming you don't invite any super villains with double-face coins over to your house, let's say two siblings decide to play the classic example of a Nash mixed-strategy equilibrium: matching pennies. In this game, Jim and his brother Tim both have a penny and they will reveal the penny, either heads up or tails, to one another. Each time both show matching pennies of heads or tails, Jim wins $10. Each time you show mismatching pennies, Tim wins $10. We can illustrate this in a matrix as follows ...

Jim↓/Tim →	**Heads**	**Tails**
Heads	10; -10	-10; 10
Tails	-10; 10	10; -10

As you can see in the matrix, if both Jim and Tim get either heads or tails, Jim wins $10 and Tim loses $10. If Jim shows heads and Tim shows tails or vice versa, Tim gets $10 and Jim loses $10.

This is what we call a zero-sum game. Why? Because the result of the two players in each box adds up to 0. Let's do the math: $10 + (-10) = 10-10 = 0$. Or $-10 + 10 = 0$. Jim and Tim have diametrically opposed interests in this game.

Zero-sum games are games that have a finite number of resources. Think back to the example of a large pizza used in the introduction. One person got a tiny piece and one person overindulged. There is only so much pizza to go around.

Poker is considered to be one of the most perfect zero-sum games ever created, and the biggest and best example of the game is the World Series of Poker (WSOP). The first WSOP was held in 1970 and included seven of the best poker players to play in the Las Vegas reunion-style game for a $5,000 buy-in. The recordkeeping wasn't great that year, but if you do anything twice in a row, it becomes a tradition. In 1971, the WSOP was born with six entrants and a $5,000 buy-in, making the total pot worth $30,000. Today, the WSOP winner wins millions of dollars, depending on the number of participants, each paying a $10,000 buy-in. The reason poker is looked as an example of a zero-sum game is because everyone playing will come in with some money, and over the course of the game, each player will strategize and make bets based on the hand they are dealt, but only one player is going to walk away with the money used to buy in.

That person "won" and played the best game of strategy against his opponents.

Non-zero-sum games have results greater or less than 0. This means the resources are not constant. If you think about it in terms of the stock market, you can make trades and you might make good trades and bad trades. Those trades might make you some money and you might lose some. Yet you can continue to make trades on that stock market and even make trades on a margin account. You could make purchases in cryptocurrency that yield you a windfall and find yourself flush with cash, or you could find yourself in a perilous and dangerous financial situation you don't know how to work your way out of because you purchased pump-and-dump stock and owe thousands of dollars you now have to pay back to a broker immediately. One player's loss doesn't necessarily correspond to the loss of another.

But back to our matching pennies:

Jim↓/Tim →	**Heads**	**Tails**
Heads	10; -10	-10; 10
Tails	-10; 10	10; -10

Why doesn't it have a pure-strategy Nash equilibrium? Because if Tim knows Jim is going to play heads or tails, he will simply play the opposite side to win the $10. Let's prove this. Could the heads-heads choice be a Nash equilibrium? No. Tim can deviate for a better outcome playing tails if he knows that Jim will play heads. If Jim plays heads and Tim plays tails, Jim will want to play tails too. If we examine each box, we will see that in each of them, one player can have a better outcome if he changes strategies.

The Nash theorem states that there has to be at least one Nash equilibrium in all games with a finite number of moves.[xliv] Since this game has no pure strategies, there must be another kind of strategy. We call these mixed strategies.

Mixed strategies are a probability distribution over one or more pure strategies.[xlv] This means that the players will select their choice randomly, in equilibrium. Each player in the matching penny game will be choosing heads or tails blindly. In addition, if the mixtures are best responses for all players, the set of strategies is a mixed-strategy Nash equilibrium.[xlvi] In the penny matching game, if both Jim and Tim are selecting heads or tails at random (flip the coins, in other words), they could assume that they will win about fifty percent of the time and lose fifty percent of the time. This is the definition of a mutual best response.

Let's assume Tim decides to flip the coin, getting heads half the time and tails half the time. Jim sticks to playing heads only. In this case, Jim will only win half of the time, as well.

Jim↓ / Tim →	**Heads (0.5)**	**Tails (0.5)**
Heads	10; -10	-10; 10

The opposite is also true. If Tim flips the coin and half of the time gets heads and half of the time tails, and Jim plays only tails, Jim will still only win half of the time.

Jim↓/Tim →	**Heads (0.5)**	**Tails (0.5)**
Tails	-10; 10	10; -10

It doesn't matter what Jim and Tim do—they can't change their outcome. They will each win or lose half of the time.

Jim↓/Tim →	**Heads (0.5)**	**Tails (0.5)**
Heads (0.5)	10; -10	-10; 10
Tails (0.5)	-10; 10	10; -10

There is no way that either of them will do better if they change their strategy. This is what a mutual best response is. Tim is best-responding to what Jim is doing. He can't secure a win or a loss, so he might just flip the coin and see what happens. The same stands true in Jim's case, too. Neither can gain more by changing the coin flipping strategy. Therefore, we found a Nash equilibrium.

Not every game is so easy to guess as just flipping a coin. Let's give different weight to each outcome in the matching pennies game.

Jim↓/Tim→	**Heads**	**Tails**
Heads	20; -20	-30; 30
Tails	-10; 10	0; 0

Here, if both Jim and Tim flip tails, there is no consequence. If both flip heads, Jim gains 20 and Tim loses 20. If Jim plays heads and Tim plays tails, Jim loses 10 and Tim gets 10. If Jim plays tails and Tim plays heads, Jim loses 30 and Tim wins 30. This payoff matrix doesn't reveal an obvious Nash equilibrium. In this case, the two players are not indifferent to the outcome of the coin flipping. To solve games like this, we will need to use some—easy—math. It is an algorithm for mixed strategies. We will learn about it more in the next chapter.

The Reality of Matching Pennies

In the real world, playing the matching pennies game likely wouldn't go so smoothly for a variety of reasons. One of the primary ones is that, as humans, we are not great at randomization. Imagine playing the matching penny game for more than five minutes. It would probably get old fast even if you were earning money. The Monte Carlo fallacy is when we falsely believe that if an event occurred more frequently than normal in the past, it is less likely to happen in the future.[xlvii] So, say our opponent randomly flipped five heads one after the other, we will likely assume that he will play tails next and adjust our strategy accordingly. The other issue is that, as humans, we naturally look for patterns and try to detect what option our opponent will choose.

We've played some games thus far, and we will be playing quite a few more. Next, we'll look closer at the mixed-strategy algorithm in the Nash equilibrium.

Key Takeaways
- ➤ If a game doesn't used a pure-strategy Nash equilibrium, look for a mixed-strategy Nash equilibrium, as we know all finite games have at least one Nash equilibrium.

Chapter 4: Mixed-Strategy Algorithm[xlviii]

In 1950, John Nash published his famous paper "Equilibrium Points in n-person Games." He defined his theorem as a set of strategies in a game where no player has any incentive to change their strategy based on the actions of other players. As we've discussed previously, Nash equilibria can be divided into two types: pure strategies and mixed strategies. Pure-strategy Nash equilibrium occurs when all players in a game are using pure strategies. Pure strategies are used when a player knows what move they will make in any given situation throughout game play.

The second type of Nash equilibrium, discussed in Chapter 3, is the mixed-strategy Nash equilibrium. In its simplest terms, mixed strategy is the application of probability to any pure strategy. When a player uses a mixed strategy, it is because the game won't permit the use of a pure strategy. The player has no other choice but to select a pure strategy at random, assign probability to it, and hope for the highest "expected payoff." The payoff is anticipated because it is based on the probability on the specific strategy working effectively. Mixed-strategy Nash equilibrium happens when at least one of the players uses mixed strategy. Not every player has to.

Let's go back to our matching pennies game from the last chapter.

Jim↓/Tim →	Heads (0.5)	Tails (0.5)
Heads (0.5)	10; -10	-10; 10
Tails (0.5)	-10; 10	10; -10

In this matrix, we could find no pure strategy Nash equilibria. We also established that if both players decided to flip their coins, they would both be indifferent about getting heads or tails. Neither of them would have done better with a different strategy—and this fact proves that there is a Nash equilibrium in the game. Just not a pure one. So, the act of coin flipping was the mixed strategy. We also discussed that not all games have such an easily detectable payoff structure.

Jim↓/Tim →	**Left**	**Right**
Up	20; -20	-30; 30

| **Down** | -10; 10 | 0; 0 |

In this game, I changed the heads and tails for left-right, up-down strategies. It makes the game easier to follow. In this game, the outcome is not so obvious. To solve this game, we are going to use an algorithm for mixed strategies to see which mixed strategies for each player makes the other player indifferent.

First, we need to find a mixed strategy for Jim. He will play down in some cases, and up in others. We need to define a strategy for Jim where Tim will be indifferent to selecting left or right. If Tim gets the same payoff if he selects left or right, it won't matter to him which way to go. He can choose at random between left or right. And if Tim's random choice makes Jim indifferent between choosing up or down, it means Jim is satisfied with his original mixed strategy. Thus, neither of them will have any incentive to change their strategy. In other words, we will have a mixed-strategy Nash equilibrium.

I know, this sounded like Albert Einstein talking in Dothraki. But bear with me. Everything will make much more sense as we delve into the solution of this problem.

Let's start with Jim's mixed strategy. We want to create a mixed strategy that makes Tim's *expected utility* for selecting left, his payoff if he goes left as a pure strategy, equal to his expected utility (or payoff) if he goes right as a pure strategy.[xlix] We will mark expected utility as EU, left as L, and right as R. And we will illustrate everything we talked about in this paragraph as:

$$EU_L = EU_R$$

Remember that mixed strategies are based on a probability. Each part of the equation is a function of a mixed strategy, and represented in the following way:

$$EU_L = f(\sigma_U)$$
$$EU_R = f(\sigma_U)$$

In these two equations, f represents the function and σ_U (σ = sigma, U = up) shows the probability that Jim plays up. Wow, we have three equations with three unknowns!

Jim↓/Tim →	Left	Right
Up	20; -20	-30; 30
Down	-10; 10	0; 0

If Tim goes left, his payoff depends on the decision Jim makes—whether he will move up or down. If Jim goes up, Tim gets -20. If Jim goes down, Tim gets 10. This is what $EU_L = f(\sigma_U)$ shows. The same stands true if Tim decides to go right. If Jim goes up, Tim gets 30. If Jim goes down, Tim gets 0. This is what this equation says: $EU_R = f(\sigma_U)$.

Jim ↓ / Tim →	Left
Up	20; **-20**
Down	-10; **10**

To solve for $EU_L = f(\sigma_U)$, we need to look at what Jim's expected utility for going left is as a function of the mixed strategy σ_U. Tim will get -20 for a percentage of the time and 10 for some percentage of the time (look at the bolded, underlined outcomes for Tim). The equation for the expected utility if Tim chooses to play left looks like this:

$$EU_L = \sigma_U(-20) + (1 - \sigma_U)(10)$$

To solve for Tim selecting left, sigma up is multiplied by -20 and added to 1 minus sigma up that has been multiplied by 10. Let's zoom in on each element of the equation above, one by one. Sigma up (σ_U) shows the probability of Jim playing up. That percentage of the time, Tim's outcome will be -20. We need to add this outcome to what happens the rest of the time. $1 - \sigma_U$ is the probability that Jim plays down. In this case, Tim is getting 10 as a payoff. So, we multiply the percentage of the time when Jim plays down $(1-\sigma_U)$ with Tim's payoff: 10.

Jim ↓ / Tim →	Right
Up	-30; **30**
Down	0; **0**

We have to apply the same logic to the case when Tim chooses to move right. Let's write the equation for the expected utility of Tim going right:

$$EU_R = \sigma_U(30) + (1 - \sigma_U)(0)$$

What does this mean? This is the expected utility of going right as a function of sigma up (σ_U). Again, Jim goes up in some cases and then Tim's payoff is 30. At occasions, Jim will go down, and then Tim's payoff is 0. (See the bolded, underlined numbers of Tim.) As we see in the equation, sigma up (σ_U) shows the probability of Jim playing up. That percentage of the time, Tim's outcome will be 30. We need to add this outcome to what happens the rest of the time. $1 - \sigma_U$ is the probability that Jim plays down. In this case, Tim is getting 0 as a payoff. So, we multiply the percentage of the time when Jim plays down ($1-\sigma_U$) with Tim's payoff: 0.

Now, we know from the very first equation that $EU_L = EU_R$. Let's set both our equations up this way and work out the algebra:

$$\sigma_U(-20) + (1 - \sigma_U)(10) = \sigma_U(30) + (1 - \sigma_U)(0)$$

How do we solve this monster? I will guide you through it step by step. It is (not so) basic algebra.

Step 1: Simplify both sides of the equation.

$$\sigma_U(-20) + (1-\sigma_U)(10) = \sigma_U(30) + (1-\sigma_U)(0)$$

$$\sigma_U(-20) + (1)(10) + (-\sigma_U)(10) = \sigma_U(30) + (1-\sigma_U)(0) \text{ (Distribute)}$$

$$-20\sigma_U + 10 + -10\sigma_U = 30\sigma_U + 0$$

$$[\mathbf{-20\sigma_U + (-10\sigma_U)}] + (10) = \mathbf{(30\sigma_U)} + (0) \text{ (Combine Like Terms)}$$

$$\mathbf{-30\sigma_U} + 10 = \mathbf{30\sigma_U}$$

$$-30\sigma_U + 10 = 30\sigma_U$$

Step 2: Subtract $30\sigma_U$ from both sides.

$$-30\sigma_U + 10 - \mathbf{30\sigma_U} = 30\sigma_U - \mathbf{30\sigma_U}$$

$$-60\sigma_U + 10 = 0$$

Step 3: Subtract 10 from both sides.

$$-60\sigma_U + 10 - \mathbf{10} = 0 - \mathbf{10}$$

$$-60\sigma_U = -10$$

Step 4: Divide both sides by -60.

$$-60\sigma_U \ / \mathbf{-60} = -10 \ / \mathbf{-60}$$

$$\sigma_U = 1/6$$

Answer:

$\sigma_U = 1/6$

What this tells us is that when Jim plays up 1/6 of the time and down 5/6 of the time, then it doesn't matter if Tim plays left or right. Tim will have the same expected utility. Now we can follow the same steps as above to find a strategy that will make Jim indifferent to Tim's decision to move left or right. We ask, what's Jim's expected utility for up or down when those two things are equal? Let's illustrate it like this:

$$EU_U = EU_D$$

The expected utility of playing up is a function of Tim's choice of playing left.
$EU_U = f(\sigma_L)$

Similarly, the expected utility of playing down is a function of Tim's decision of playing left.

$$EU_D = f(\sigma_L)$$

Let's bring everything together:

Jim↓/Tim →	Left	Right
Up	**20**; -20	**-30**; 30

Jim's expected utility to play up is 20 if Tim moves left and -30 if Tim moves right. See the equation below:

$$EU_U = \sigma_L(20) + (1 - \sigma_L)(-30)$$

Jim↓/Tim →	Left	Right
Down	**-10**; 10	**0**; 0

Jim's expected utility to play down is -10 if Tim moves left and 0 if Tim moves right.

$$EU_D = \sigma_L (-10) + (1- \sigma_L) (0)$$

As $EU_U = EU_D$, we can write our equation as follows:

$$\sigma_L (20) + (1 - \sigma_L) (-30) = \sigma_L (-10) + (1- \sigma_L) (0)$$

Let's solve this equation step-by-step.

Step 1: Simplify both sides of the equation.

$$\sigma_L (20) + (1-\sigma_L)(-30) = \sigma_L (-10) + (1-\sigma_L)(0)$$

$$\sigma_L\,(20) + (1)(-30) + (-\sigma_L)(-30) =$$
$$\sigma_L\,(-10) + (1-\sigma_L)(0) \text{ (Distribute)}$$

$$20\sigma_L + -30 + 30\sigma_L = -10\sigma_L + 0$$

$$(\mathbf{20\sigma_L + 30\sigma_L}) + (-30) = (\mathbf{-10\sigma_L}) +$$
(0) (Combine Like Terms)

$$[\mathbf{50\sigma_L} + \mathit{(-30)}] = \mathbf{-10\sigma_L}$$

$$50\sigma_L + -30 = -10\sigma_L$$

Step 2: Add $10\sigma_L$ to both sides.

$$50\sigma_L + -30 + \mathbf{10\sigma_L} = -10\sigma_L + \mathbf{10\sigma_L}$$

$$60\sigma_L + -30 = 0$$

Step 3: Add 30 to both sides.

$60\sigma_L + -30 + \mathbf{30} = 0 + \mathbf{30}$

$60\sigma_L = 30$

Step 4: Divide both sides by 60.

$60\sigma_L / \mathbf{60} = 30 / \mathbf{60}$

$\sigma_L = 1/2$

Answer:
$\sigma_L = 1/2$

The end result is that when Tim has the same expected utility and plays left with the probability 1/2 of time and right with a probability 1/2 of the time, Jim is indifferent when he plays up or down.

Let's fill our game matrix with all the information we got. So, in this game, the mixed-strategy Nash equilibrium is when Jim plays up 1/6 of the time and Tim plays left 1/2 of the time.

$$\sigma_U = 1/6 \text{ and } \sigma_L = 1/2$$

Jim ↓ / Tim →	**Left 1/2**	**Right 1/2**
Up 1/6	20; -20	-30; 30
Down 5/6	-10; 10	0; 0

As long as Tim and Jim stick to these strategies, neither one can change what they do and expect better results. Why? Because Jim's expected utility for playing up and down are the same, and so is Tim's for playing left or right.

Mixed Strategy and Soccer Penalty Kicks

I am a big soccer fan. I don't have a particular favorite team; I just enjoy watching the game. Especially the World Cup or the Euro Cup. The story I want to tell you happened in the Euro 2004 quarterfinals, in the match between Portugal and England. The two teams couldn't get a conclusive score in game time and the two short game extensions, so it all came down to penalty kicks. David Beckham, the legend of England, was the first penalty kicker. He messed up this kick so gloriously that I still remember him because of this penalty rather than his general brilliance.

Not only was this penalty kick terribly executed, but David Beckham had missed two other penalty kicks during this championship. His last miss was considered one of the worst penalty misses in football history, and led to England's elimination from the tournament.[1]

Arguably, penalty kicks are some of the most suspenseful moments in a soccer game. But they are also a great example of game theory's mixed strategy. Let's see how:

Kicker↓/ Goalie →	**Lean Left**	**Lean Right**
Kick Left	0; 0	2; -2
Kick Right	1; -1	0; 0

The matrix above illustrates the payoff of a soccer game. Each time a kicker and a goalie face each other *The Good, the Bad, and the Ugly*-style, the kicker has to decide whether to kick left or right and the goalie has to decide whether to block left or right. If the kicker kicks left and the goalie leans left, the goalie will save the ball, so both parties get 0. The same thing happens if the kicker kicks right and the goalie leans right.

If, however, the kicker kicks right and the goalie leans left, that gives one point for the kicker and minus one for the goalie's team. Additionally, kickers usually are right-footed, which means they can shoot much better towards left. Thus, if the kicker kicks left and the goalie leans right, the kicker gets two and the goalie minus two.

This game has no pure-strategy Nash equilibrium. Why? Because each player would deviate from any of the strategies. For example, kick left, lean left is not an equilibrium because the kicker would deviate to kick right and increase the payoff from 0 to 1.[li] In the kick left, lean right scenario, the goalie would want to change strategies, leaning rather left as 0 is a better outcome than -2. In the case of kick right, lean left, again, the goalie would want to shift strategies and lean right as 0 is greater than -1. And finally, in the kick right, lean right scenario, the kicker would want to change strategies and kick left as 2 is bigger than 0.

Now we know that not having a pure Nash equilibrium doesn't mean that there is no Nash equilibrium at all. We just have to use our mixed strategy algorithm to find it.

We can deduct the kicker's mixed-strategy equilibrium from knowing that, unless the payoffs from kicking left or kicking right are equal, he will deviate from randomizing. Let's recall our equations.

We know that the expected utility of leaning left ($_{LL}$) has to be equal to the expected utility of leaning right ($_{LR}$).

$$EU_{LL} = EU_{LR}$$

We also know that each part of the equation is a function of a mixed strategy, and represented in the following way:

$$EU_{LL} = f(\sigma_{KL})$$
$$EU_{LR} = f(\sigma_{KL})$$

In these two equations, f represents the function and σ_{KL} (σ = sigma, KL = kicking left) shows the probability that the kicker kicks left. If the kicker kicks left, his payoff depends on the decision the goalie makes—whether he will lean left or right. If the kicker goes left, the goalie gets 0. If the kicker goes right, the goalie gets -1. This is what $EU_{LL} = f(\sigma_{KL})$ shows. The same stands true if the kicker decides to go right. If the kicker goes right, the goalie gets -2. If the kicker goes right, the goalie gets 0. This is what this equation says: $EU_{LR} = f(\sigma_{KL})$.

Kicker ↓ / Goalie →	**Lean Left**
Kick Left	0; **0**
Kick Right	1; **-1**

To solve for $EU_{LL} = f(\sigma_{KL})$, we need to look at what the kicker's expected utility for kicking left is as a function of the mixed strategy σ_{KL}. The goalie will get 0 for a percentage of the time and -1 for some percentage of the time (look at the bolded, underlined outcomes for the goalie). The equation for the expected utility if the goalie chooses to lean left looks like this:

$$EU_{LL} = \sigma_{KL}(0) + (1 - \sigma_{KL})(-1)$$

To solve for the goalie leaning left, sigma $_{KL}$ is multiplied by 0 added to 1 minus sigma KL that has been multiplied by -1. Let's zoom in on each element of the equation above, one by one. Sigma $_{KL}$ (σ_{KL}) shows the probability of the kicker playing left. That percentage of the time, the goalie's outcome will be 0. We need to add this outcome to what happens the rest of the time. 1 - σ_{KL} is the probability that the kicker plays right. In this case, the goalie is getting -1 as a payoff. So, we multiply the percentage of the time when the kicker plays right $(1-\sigma_{KL})$ with the goalie's payoff: -1.

Kicker ↓ / Goalie →	**Lean Right**
Kick Left	2; **-2**
Kick Right	0; **0**

We have to apply the same logic to the case when the goalie chooses to lean right. Let's write the equation for the expected utility of the kicker kicking right:

$$EU_{LR} = \sigma_{KL}(-2) + (1 - \sigma_{KL})(0)$$

What does this mean? This is the expected utility of kicking right as a function of sigma $_{KL}$ (σ_{KL}). Again, the kicker goes left in some cases and then the goalie's payoff is

-2. At occasions, the kicker will kick right, and then the goalie's payoff is 0. (See the bolded, underlined numbers of the goalie.) As we see in the equation, sigma $_{KL}$ (σ_{KL}) shows the probability of the kicker kicking left. That percentage of the time, the goalie's outcome will be -2. We need to add this outcome to what happens the rest of the time. 1 - σ_{KL} is the probability that the kicker plays right. In this case, the goalie is getting 0 as a payoff. So, we multiply the percentage of the time when the kicker plays right (1 - σ_{KL}) with the goalie's payoff: 0.

As the expected utility of leaning left and right is equal ($EU_{LL} = EU_{LR}$), we can write down our equation:

$$\sigma_{KL}(0) + (1 - \sigma_{KL})(-1) = \sigma_{KL}(-2) + (1 - \sigma_{KL})(0)$$

Let's solve it!

Step 1: Simplify both sides of the equation.

$$\sigma_{KL}(0) + (1-\sigma_{KL})(-1) = \sigma_{KL}(-2) + (1 - \sigma_{KL})(0)$$

$$\sigma_{KL}(0) + (1)(-1) + (-\sigma_{KL})(-1) = \sigma_{KL}(-2) + (1 - \sigma_{KL})(0) \text{ (Distribute)}$$

$$0 + -1 + \sigma_{KL} = -2\sigma_{KL} + 0$$

$$(\sigma_{KL}) + (0+-1) = (-2\sigma_{KL}) + (0) \text{ (Combine Like Terms)}$$

$$\sigma_{KL} + -1 = -2\sigma_{KL}$$

$$\sigma_{KL} + -1 = -2\sigma_{KL}$$

Step 2: Add $2\sigma_{KL}$ to both sides.

$\sigma_{KL} + -1 + 2\sigma_{KL} = -2\sigma_{KL} + 2\sigma_{KL}$

$3\sigma_{KL} + -1 = 0$

Step 3: Add 1 to both sides.

$3\sigma_{KL} + -1 + 1 = 0 + 1$

$3\sigma_{KL} = 1$

Step 4: Divide both sides by 3.

$3\sigma_{KL} / 3 = 1 / 3$

$\sigma_{KL} = 1/3$

Answer:

$\sigma_{KL} = 1/3$

What this tells us is that when the kicker plays left 1/3 of the time and right 2/3 of the time, then it doesn't matter if the goalie leans left or right. The goalie will have the same expected utility. Now we can follow the same steps as above to find a strategy that will make the kicker indifferent to the goalie's decision to lean left or right. We ask, what's the kicker's expected utility for kicking left or right when those two things are equal? Let's illustrate it like this:

$$EU_{KL} = EU_{KR}$$

The expected utility of kicking left is a function of the goalie's choice of leaning left.

$$EU_{KL} = f(\sigma_{LL})$$

Similarly, the expected utility of kicking right is a function of the goalie's decision of leaning left.

$$EU_{KR} = f(\sigma_{LL})$$

Let's bring everything together:

Kicker ↓ / Goalie →	**Lean Left**	**Lean Right**
Kick Left	**0**; 0	**2**; -2

The kicker's expected utility to play left is 0 if the goalie moves left and 2 if the goalie moves right. See the equation below:

$$EU_{KL} = \sigma_{LL}(0) + (1 - \sigma_{LL})(2)$$

Kicker ↓ / Goalie →	**Lean Left**	**Lean Right**
Kick Right	1; -1	0; 0

The kicker's expected utility to play right is 1 if the goalie moves left and 0 if the goalie moves right.

$$EU_{KR} = \sigma_{LL}(1) + (1 - \sigma_{LL})(0)$$

As $EU_{KL} = EU_{KR}$, we can write our equation as follows:

$$\sigma_{LL}(0) + (1 - \sigma_{LL})(2) = \sigma_{LL}(1) + (1 - \sigma_{LL})(0)$$

Let's solve your equation step-by-step.

Step 1: Simplify both sides of the equation.

$$\sigma_{LL}(0) + (1 - \sigma_{LL})(2) = \sigma_{LL}(1) + (1 - \sigma_{LL})(0)$$

$$\sigma_{LL}(0) + (1)(2) + (-\sigma_{LL})(2) = \sigma_{LL}(1) + (1 - \sigma_{LL})(0) \text{ (Distribute)}$$

$$0 + 2 + -2\sigma_{LL} = \sigma_{LL} + 0$$

$$(-2\sigma_{LL}) + (0+2) = (\sigma_{LL}) + (0) \text{ (Combine Like Terms)}$$

$$-2\sigma_{LL} + 2 = \sigma_{LL}$$

Step 2: Subtract σ from both sides.

$$-2\sigma_{LL} + 2 - \sigma_{LL} = \sigma_{LL} - \sigma_{LL}$$

$$-3\sigma_{LL} + 2 = 0$$

Step 3: Subtract 2 from both sides.

$$-3\sigma_{LL} + 2 - 2 = 0 - 2$$

$$-3\sigma_{LL} = -2$$

Step 4: Divide both sides by -3.

$$-3\sigma_{LL} / -3 = -2 / -3$$

$$\sigma_{LL} = 2/3$$

Answer:
$\boldsymbol{\sigma_{LL} = 2/3}$

The end result is that when the goalie has the same expected utility and leans left with the probability 2/3 of time and right with a probability 1/3 of the time, the kicker is indifferent when he kicks left or right.

Let's fill our game matrix with all the information we've got. So, in this game, the mixed-strategy Nash equilibrium is when the kicker plays left 1/3 of the time and the goalie leans left 2/3 of the time.

$$\sigma_{KL} = 1/3 \text{ and } \sigma_{LL} = 2/3$$

Kicker↓/Goalie→	**Lean Left 2/3**	**Lean Right 1/3**
Kick Left 1/3	0; 0	2; -2
Kick Right 2/3	1; -1	0; 0

As long as the kicker and the goalie stick to these strategies, neither one can change what they do and expect better results. Why? Because the kicker's expected utility for kicking left and right are the same, and so is the goalie's for leaning left or right.

Mixed-Strategy Nash Equilibrium in Real Life

We just saw how the mixed-strategy Nash equilibrium can play out in soccer. But it generally benefits any competitive game. Have you ever watched a professional tennis game? You can observe that top players mix their shots all the time. If Nadal always played the same shot, then Federer could quickly learn to anticipate the move and position to return the shot in a way Nadal wouldn't expect. If Federer responded the same way to Nadal's identical shots, Nadal would learn to anticipate Federer's response, too. The game would get boring, predictable, and assuming none of them got tired, the shoot-throughs would go on in perpetuity.

Mixed-strategy Nash equilibrium is used in poker, too. Sometimes players bluff or pretend to bluff by sandbagging a good hand. Predictability weakens their hand because their opponents can anticipate their actions and position themselves accordingly. Just like Sun Tzu said, the element of surprise and being inconsistent can be advantageous. This is an unwritten law in the playbook of sports, business, and war. Surprise calls for a mixed strategy almost by definition.

Another example of mixed-strategy Nash equilibrium is a diversified portfolio on the stock market. If one had perfect information about the yield of every stock in the stock market, their best response would be to choose the stock with the highest yield and put all their money into it. That would be a pure Nash strategy. But we don't have perfect, complete information about the stock market. Most of us mix our strategy, creating unique types of mixed-strategy Nash equilibria.

Chapter 5: Pure and Mixed Nash Equilibria: Mixed

In this chapter, we will learn more about simultaneous games. Remember, in these types of games, both players apply their strategies at the same time. Many fields use simultaneous games from the military to find the best way to act.

We don't need to think in terms of the battle tactics of Julius Caesar to picture simultaneous games. The simplest example is probably a game we all played at some point in our lives: rock, paper, scissors. In this game we have *complete information*. We know what we need to do to win or lose:

paper beats rock, rock beats scissors, scissors beats paper. This knowledge gives some utility; we may strategize which "weapon" to choose and how frequently. We are aware that other players have the same information. The rules of the game and the payoffs are known by everybody.

Simultaneous games usually use the normal form. The game is illustrated in a matrix. We assume that our opponent is a rational decision-maker and will make the best moves to maximize their benefit. When both us and our opponent reach a point of equilibrium, following rational choices, where we have no incentive to change our strategy, we reach a Nash equilibrium.

Do the paragraphs above help you recall something you learned already in this book? Yes, the prisoner's dilemma. That was the perfect example of a simultaneous game with complete information, in normal form. To solve this game, we worked on finding a strictly dominant strategy (betray). This always offers a better individual outcome to the players than the alternative: to stay silent. We, therefore, discounted the *strictly dominated* strategies. This rational approach helps eliminate undesirable strategies in a systematic way.

But did the prisoner's dilemma reach a socially optimal result? The social optimum would have been for both players to stay silent, get a shorter punishment, and get out within a year instead of eight. In game theory, social inefficiencies are common when dominant strategies don't serve the common good.[lii]

Let's see an example where players want to cooperate, there is no strictly dominant strategy, and no pure Nash equilibrium.

The Battle of the Sexes

In this chaotic and fast-paced world, there is something intangible that connects us to each other and gives us the sense to introduce order and meaning into our lives. This is our ability to communicate with each other. Most humans are by nature gregarious and social beings. But, what if two people don't have any means of reaching each other? This happened with Jim and his girlfriend, Kim. They wanted to go on a date but couldn't talk on their mobile phones.

Jim and Kim are dating for three years. Each Friday night, they plan on going out for a date to spend quality time together. Both of them have their own preferences about where to go. Kim prefers an evening attending a lively dance class. Jim's preference is to watch an adrenaline-pumping, action-packed rugby game. This week, they couldn't agree where to go, just that they should meet at seven. They wanted to decide later in the day whether to go to dance or watch rugby. But during the day, Kim's phone battery died and now they have no way to know which location they should show up at at seven. Jim and Kim want to be together either at a dance class or at rugby. It wouldn't classify as a date for Jim to be alone watching rugby and for Kim to attend her dance class alone, but simultaneously.

They could end up in different places if they are not able to communicate with each other before they leave their workplaces to rendezvous on their date. The only way they would both go to the same venue is if they decide beforehand where to go. In other words, we are dealing with a game with incomplete information.

We play many games in our daily lives. Most of them are played sequentially; players or teams play one after the other. Consider chess. We see what move our opponent has played and, when it is our turn, we make our move, and vice-versa. Some games are played differently. Two players or teams make their move concurrently, not knowing what move the other player is making. Jim and Kim are in such a game. It is a simultaneous game that game theorists call the battle of sexes.

The battle of sexes is a classic *coordination game*. Jim and Kim wanted to do different things on their date. But their greater desire was to be together, be it at a dance class or a rugby game. To achieve togetherness, one has to sacrifice their personal choice for the other.

Jim and Kim have four possible outcome combinations:

1. Kim at dance class, Jim at dance class.
2. Kim at dance class, Jim at rugby game.
3. Kim at rugby game, Jim at dance class.

4. Kim at rugby game, Jim at rugby game.

Out of the four outcomes, the couple ends up together only in scenarios 1 and 4. Let's illustrate this in a game matrix.

Jim↓/Kim →	**Dance**	**Rugby**
Dance	1; 2	0; 0
Rugby	0; 0	2; 1

In outcome dance-dance, Kim is fully satisfied as she has her preference and also has Jim with her. The outcome rugby-rugby makes Jim very happy as he has his preference and also has Kim with him. Both in outcomes dance-rugby or rugby-dance, one member of the couple is at their preferred place but they are not together.

They don't want to be alone so they just go home, unhappy.

Let's try to find some pure-strategy Nash equilibria. Remember, it is a set of strategies, one for each player, where no player has the incentive to change their strategy given what the other players are doing.[liii]

The incentive for both Jim and Kim is to be together. Knowing this, let's see each option.

Is dance-dance a Nash equilibrium? Yes.

Jim ↓ / Kim →	**Dance**
Dance	1; 2

We see that Kim is most satisfied as she is at the dance class and she also has Jim with her. Jim is happy just being with her even though he is not at his preferred rugby. Kim wouldn't want to deviate from this outcome and go to the rugby game as 2 is bigger than 0. Jim also doesn't want to change strategies and go to rugby as 1 is greater than 0. This outcome is a Nash equilibrium. As we learned earlier in Chapter 2 of this book, this is a pure strategy adopted by both players who have no incentive to change strategies and hope for a better outcome.

What about when Kim goes to dance, and Jim goes to rugby? Is this a Nash equilibrium?

Jim ↓ / Kim →	**Dance**
Rugby	0; 0

No, it is not. If Kim changed strategies and went to rugby, she would be with Jim and get 1. Even if this is her lesser preferred activity, they would still be together. What about Jim? He would also fare better if he switched to go dancing. He would meet Kim and he would get 1 instead of 0.

The opposite option isn't a Nash equilibrium either.

Jim ↓ / Kim →	**Rugby**
Dance	0; 0

If either of them deviates from their strategies, they will be better off. If Jim switches to rugby, he gets 2 instead of 0. Also, if Kim goes to dance instead of the rugby game, she gets 2 instead of 0.

The last unexamined option is when Kim and Jim both go to rugby.

Jim ↓ / Kim →	**Rugby**
Rugby	2; 1

This option, just like the dance-dance one, is a Nash equilibrium. Why? Because neither of them would be better off if they switched strategies. If Kim went dancing instead of the rugby game, she would get 0, not 1. And if Jim chose the dance class, he would trade his 2 for a 0.

Battle of the sexes has two pure-strategy Nash equilibria. Jim and Kim wanted to be together doing what the other partner preferred while sacrificing what they personally would have liked to do. So, we can see the incentive for cooperation. But we can also see that their preference differs. Coordination, in this case, is not very clear. We can't be sure if they will end up in the dance-dance outcome or the rugby-rugby outcome. What can we do?

Just like in the previous chapter, when we can't decide what the best move using pure strategies is, we should take a look at mixed strategies. And yes, we need to use our mixed-strategy algorithm to do it.

To make the distinctions easier, I renamed the strategies as follows:

Jim↓/Kim →	**Dance (KD)**	**Rugby (KR)**
Dance (JD)	1; 2	0; 0
Rugby (JR)	0; 0	2; 1

KD stands for Kim Dance and KR stands for Kim Rugby. JD stands for Jim Dance and JR stands for Jim Rugby. Now let's see the algorithm, first exploring Jim's mixed strategy.

We ask what the expected utility of KD (Kim Dance) and expected utility of KR (Kim Rugby) is as a function of Jim's mixed strategy. Remember, they have to be equal.

$$EU_{KD} = EU_{KR}$$

$$EU_{KD} = f(\sigma_{JD})$$

$$EU_{KR} = f(\sigma_{JD})$$

First, we will calculate Kim's dancing strategy as a function of Jim's strategy sigma $_{JD}$ (f(σ_{JD}). (See the gray box and bolded, underlined letters.)

Jim↓/Kim →	**Dance (KD)**	**Rugby (KR)**
Dance (JD)	1; **2**	0; 0
Rugby (JR)	0; **0**	2; 1

Kim will get 2 some percentage of the time (when Jim goes to dance, $_{JD}$) and 0 some percentage of the time (when Jim goes to the rugby game, $_{JR}$).

$$EU_{KD} = \sigma_{JD}(2) + (1 - \sigma_{JD})(0)$$

Sigma $_{JD}$ times 2 ($\sigma_{JD}(2)$) is the probability that Jim will go to dance and 1 minus sigma $_{JD}$ times 0 ($1 - \sigma_{JD}$) (0) is the probability of Jim going to rugby. And this collectively is Kim's payoff for choosing to go to dance.

Let's see the other side and calculate the expected utility of Kim going to the rugby game, $_{KR}$. Kim will get 0 some percentage of the time (when Jim goes to dance, $_{JD}$) and 1 some percentage of the time (when Jim goes to the rugby game, $_{JR}$).

$$EU_{KR} = \sigma_{JD}(0) + (1 - \sigma_{JD})(1)$$

Sigma $_{JD}$ times 0 ($\sigma_{JD}(0)$ is the probability that Jim will go to dance and 1 minus sigma $_{JD}$ times 1 ($1 - \sigma_{JD}$) (1) is the probability of Jim going to rugby. And this collectively is Kim's payoff for choosing to go to rugby.

As $EU_{KD} = EU_{KR}$, we can write our equation as follows:

$$\sigma_{JD}(2) + (1 - \sigma_{JD})(0) = \sigma_{JD}(0) + (1 - \sigma_{JD})(1)$$

Let's solve this equation step-by-step.

Step 1: Simplify both sides of the equation.

$$\sigma_{JD}(2) + (1 - \sigma_{JD})(0) = \sigma_{JD}(0) + (1 - \sigma_{JD})(1)$$

$$\sigma_{JD}(2) + (1 - \sigma_{JD})(0) = \sigma_{JD}(0) + (1)(1) + (-\sigma_{JD})(1) \text{ (Distribute)}$$

$$2\sigma_{JD} + 0 = 0 + 1 + -\sigma_{JD}$$

$$(2\sigma_{JD}) + (0) = (-\sigma_{JD}) + (0+1)$$
(Combine Like Terms)

$$2\sigma_{JD} = -\sigma_{JD} + 1$$

Step 2: Add σ_{JD} to both sides.

$$2\sigma_{JD} + \sigma_{JD} = -\sigma_{JD} + 1 + \sigma_{JD}$$
$$3\sigma_{JD} = 1$$

Step 3: Divide both sides by 3.

$$3\sigma_{JD} / 3 = 1/3$$

$\sigma_{JD} = 1/3$

Answer:
$\sigma_{JD} = 1/3$

In other words, when Jim goes to dance with a probability of 1/3 and to rugby with a probability of 2/3, Kim is indifferent whether to go to dance or rugby.

Half of our work is done. Now we need to see if Kim has a mixed strategy that leaves Jim indifferent between dance and rugby.

We ask what the expected utility of $_{JD}$ (Jim Dance) and expected utility of $_{JR}$ (Jim Rugby) is as a function of Kim's mixed strategy. Remember, they have to be equal.

$$EU_{JD} = EU_{JR}$$

$$EU_{JD} = f(\sigma_{KD})$$

$$EU_{JR} = f(\sigma_{KD})$$

Let's calculate Jim's dancing strategy as a function of Kim's strategy sigma $_{KD}$ (f(σ_{KD}). (See the gray box and bolded, underlined letters.)

Jim↓/Kim →	**Dance (KD)**	**Rugby (KR)**
Dance (JD) 1/3	**1**; 2	**0**; 0
Rugby (JR) 2/3	0; 0	2; 1

Jim will get 1 some percentage of the time (when Kim goes to dance, $_{KD}$) and 0 some percentage of the time (when Kim goes to the rugby game, $_{KR}$).

$$EU_{JD} = \sigma_{KD}(1) + (1 - \sigma_{KD})(0)$$

Sigma $_{KD}$ times 1 ($\sigma_{KD}(1)$ is the probability that Kim will go to dance and 1 minus sigma $_{KD}$ times 0 $(1 - \sigma_{KD})(0)$ is the probability of Kim going to rugby. And this collectively is Jim's payoff for choosing to go to dance.

Let's see the other side and calculate the expected utility of Jim going to the rugby game, $_{JR}$. Jim will get 0 some percentage of the time (when Kim goes to dance, $_{KD}$) and 2 some percentage of the time (when Kim goes to the rugby game, $_{KR}$).

$$EU_{JR} = \sigma_{KD}(0) + (1 - \sigma_{KD})(2)$$

Sigma $_{KD}$ times 0 ($\sigma_{KD}(0)$ is the probability that Kim will go to dance and 1 minus sigma $_{KD}$ times 2 $(1 - \sigma_{KD})(2)$ is the probability of Kim going to rugby. And this collectively is Jim's payoff for choosing to go to rugby.

As $EU_{JD} = EU_{JR}$, we can write our equation as follows:

$$\sigma_{KD}(1) + (1 - \sigma_{KD})(0) = \sigma_{KD}(0) + (1 - \sigma_{KD})(2)$$

Let's solve this equation step-by-step.

Step 1: Simplify both sides of the equation.

$$\sigma_{KD}(1) + (1 - \sigma_{KD})(0) = \sigma_{KD}(0) + (1 - \sigma_{KD})(2)$$

$$\sigma_{KD}(1) + (1 - \sigma_{KD})(0) = \sigma_{KD}(0) + (1)(2) + (-\sigma_{KD})(2) \text{ (Distribute)}$$

$$\sigma_{KD} + 0 = 0 + 2 + -2\sigma_{KD}$$

$$(\sigma_{KD}) + (0) = (-2\sigma_{KD}) + (0+2) \text{ (Combine Like Terms)}$$

$$\sigma_{KD} = -2\sigma_{KD} + 2$$

Step 2: Add $2\sigma_{KD}$ to both sides.

$$\sigma_{KD} + 2\sigma_{KD} = -2\sigma_{KD} + 2 + 2\sigma_{KD}$$

$$3\sigma_{KD} = 2$$

Step 3: Divide both sides by 3.

$3\sigma_{KD} / 3 = 2 / 3$

$\sigma_{KD} = 2/3$

Answer:

$\sigma_{KD} = 2/3$

Now we've learned that when Kim is mixing her strategy to go to dance with a probability of 2/3 and to rugby with a probability of 1/3, Jim is indifferent whether to go to dance or rugby.

Jim↓/Kim→	**Dance** **(KD) 2/3**	**Rugby** **(KR) 1/3**
Dance (JD) 1/3	1; 2	0; 0

| **Rugby (JR) 2/3** | 0; 0 | 2; 1 |

We can conclude that, in this game, there are three Nash equilibria, two pure strategies (dance-dance and rugby-rugby) and a mixed-strategy Nash equilibrium (the one we just calculated). We are not done yet. We need to find a way to compare the mixed-strategy Nash equilibrium with the pure-strategy Nash equilibria to conclude which is better.

How to Calculate Payoffs?

In the previous section, we calculated what Jim and Kim do in a mixed-strategy Nash equilibrium. We don't know the actual payoffs of their mixed strategies. Now we are going to learn how to calculate it! Let's find out the relative advantage of the pure- vs. mixed-strategy Nash equilibria.

The payoff of the two pure-strategy Nash equilibria is known:

- If both Jim and Kim go to dance, Jim gets 1 and Kim gets 2.
- If both Jim and Kim go to rugby, Jim gets 2 and Kim gets 1.

What we don't know: What the exact number for the mixed-strategy Nash equilibrium is. All we know are the probabilities:

Jim↓/Kim →	**Dance** **(KD) 2/3**	**Rugby** **(KR) 1/3**
Dance (JD) 1/3	1; 2	0; 0

| **Rugby (JR) 2/3** | 0; 0 | 2; 1 |

To make our calculation, we will need to use another algorithm.

1. First, we need to find the probability that each outcome happens in equilibrium.
2. Then, for each outcome, we need to multiply that probability with a given player's payoff.
3. Sum all these numbers together.[liv]

This calculation won't be so complicated as it sounds, I promise. How do we eat an elephant? By chopping it into bite-sized pieces. Let's start with bite number one.

1. First, we need to find the probability that each outcome happens in equilibrium.

Jim↓/Kim →	**Dance** **(KD) 2/3**	**Rugby** **(KR) 1/3**
Dance (JD) 1/3	1; 2	0; 0
Rugby (JR) 2/3	0; 0	2; 1

We've already found the probabilities of the players. See the bolded and underlined numbers. The next step is to multiply across the matrix. To see what the probability of both Jim and Kim going to dance class is, you take Jim's probability, 1/3, and multiply it with Kim's probability, 2/3.

Jim ↓ / Kim →	**Dance (KD) 2/3**
Dance (JD) 1/3	1; 2 → 2/9

1/3 (2/3) = 2/9

The result of this multiplication is 2/9. You simply write it into the dance-dance outcome's box.

We need to make the same calculation for each of the four outcomes. Once we are done, we get these results in the matrix:

1/3 (1/3) = 1/9 (JDxKR)
2/3 (2/3) = 4/9 (JRxKD)
2/3 (1/3) = 2/9 (JRxKR)

Jim↓/Kim →	**Dance** (KD) 2/3	**Rugby** (KR) 1/3
Dance (JD) 1/3	1; 2 → **2/9**	0; 0 → **1/9**
Rugby (JR) 2/3	0; 0 → **4/9**	2; 1 → **2/9**

The number fractions have to add up to 1. Why? Because probabilities by definition have to end up being 1. So, you can always check if you made your calculations right by just adding together the four numbers:

$$2/9 + 1/9 + 4/9 + 2/9$$
$$= 9/9$$
$$= 1$$

So, we did a good job here. Now we have to do what we discussed in bullet point number two.

2. For each outcome, we need to multiply that probability with a given player's payoff.

Let's start this with Jim's payoffs.

Jim↓/Kim →	**Dance (KD) 2/3**	**Rugby (KR) 1/3**
Dance (JD) 1/3	1 x **2/9**	0 x **1/9**
Rugby (JR) 2/3	0 x **4/9**	2 x **2/9**

And now we go to step 3.

3. Sum all these numbers together.

211

Let's do it:

$(1 \times 2/9) + (0 \times 1/9) + (0 \times 4/9) + (2 \times 2/9) =$

$= 2/9 + 0 + 0 + 4/9$

$= 6/9$

$= \mathbf{2/3}$

This means that Jim, in his mixed-strategy Nash equilibrium, gets 2/3 of a point.

Now we need to do the steps for Kim. The probabilities calculated in step 1 don't change.

Jim↓/Kim →	**Dance** **(KD) 2/3**	**Rugby** **(KR) 1/3**

Dance (JD) **1/3**	2 x **2/9**	0 x **1/9**
Rugby (JR) **2/3**	0 x **4/9**	1 x **2/9**

Let's sum everything:

(2 x 2/9) + (0 x 1/9) + (0x4/9) + (1 x 2/9)

= 4/9 + 0 + 0 + 2/9

= 6/9

= 2/3

This means that Kim's mixed strategy also gives her 2/3 of a point.

What does this all mean?

Let's compare the payoffs of the mixed-strategy Nash equilibrium and the payoffs of the pure-strategy Nash equilibrium.

What we can conclude is that choosing the mixed-strategy Nash equilibrium yields a worse payoff for both players. The payoff of the mixed strategy for both Jim and Kim is 2/3, which is somewhat less than 1. Thus, even if Jim or Kim end up at their least preferred activity, they are still better off going with the pure-strategy Nash equilibrium, which gives them at least 1 point.

In other words, it is always smarter for Jim to just agree at home to go to the dance class. It is smarter for him to give up his preference for Kim's choice. Similarly, Kim also fares better if she just agrees to meet at the rugby game. Her payoffs will be higher than going for the mixed strategy where they don't come to an agreement.

Overall, when you encounter a game that has both pure- and mixed-strategy Nash equilibria, it's better to calculate the mixed strategy's payoff first to see how it compares to the pure strategy's outcomes. There are games where choosing the mixed strategy yields a better payoff. And there are games, like the battle of the sexes, where the pure strategy is the better option.

Chapter 6: Strict and Weak Dominance

"Knowing the strategy of my competitors, and treating those strategies as unchangeable, would I be worse off by changing my strategy?"

If the answer is "yes" for every player involved, we are dealing with a strict Nash equilibrium. We've learned this so far in many examples, such as the prisoner's dilemma. Every strictly dominant strategy is a Nash equilibrium, but not every Nash equilibrium is a strictly dominant strategy. How so?

Let's imagine a game where, for a player, the strategy in Nash equilibrium and another strategy (not in Nash equilibrium) give the exact same payoffs. In other words, this player is indifferent between the two strategies. When this is the case, we consider that Nash equilibrium strategy *weak*.

Let's learn a little more about both of these types of dominance play out in mixed strategies.

Strict-Dominance Nash Equilibrium

Jim↓ / Tim →	Left	Right
Up	6; -2	-2; 2
Center	0; 0	0; 0
Down	-2; 4	4; -2

This is our game. As we can see, Jim has three strategies, up-center-down. Tim has two strategies, left and right.

There is strictly dominant pure strategy in this game. If you wish, it can be a good exercise for you to verify it. The practice of iterated elimination of strictly dominated pure strategies doesn't work out well here.[lv]

We know what is coming. We need to check for mixed strategies. But before we do that, let's learn a little bit about how strict dominance plays out in mixed strategies.

1. When you discover that two or more mixed strategies strictly dominate another strategy, get rid of the latter.

2. *Strictly dominated* strategies simply don't make sense to play out, no matter if they are dominated by pure or mixed strategies.

In our game, if we could detect a mixed strategy that dominates a pure strategy, we could eliminate the pure strategy first and then solve the game as we did with other mixed-strategy games.

Jim ↓ / Tim →	**Left**
Up 1/2	**6**; -2
Center	0; 0
Down 1/2	**-2**; 4

Let's assume Jim plays up with a probability of 1/2 and down with a probability of 1/2. If Tim moves left, it means that Jim gets 6 half of the time and -2 half of the time.

This will average out to be worth 1. And 1 is bigger than 0, which is what Jim would receive if he played center. The 1/2 up, 1/2 down mixed strategy is better for Jim than playing center when Tim is going left.

Jim ↓ / Tim →	**Right**
Up 1/2	**-2**; 2
Center	0; 0
Down 1/2	**4**; -2

If Tim goes right, and Jim keeps his 1/2 up, 1/2 down mixed strategy, Jim will get -2 when playing up with a probability of 1/2 and 4 when playing down with a probability of 1/2. This averages out to be 1.5. And, again, the 1.5 Jim gets from playing this mixed strategy is better than playing center for 0.

We can conclude that regardless of what Tim is doing (going left or right), Jim will be better off playing his mixed strategy rather than going center. Jim's mixed strategy thus exhausts the definition of strict dominance: He has no incentive to change his strategy no matter what Tim does.

Knowing that center is a strictly dominated strategy, we can eliminate it.

Jim↓/Tim →	**Left**	**Right**
Up	6; -2	-2; 2
Down	-2; 4	4; -2

This is how our game looks like right now. Familiar, right? At first glance, we can see that there are no pure strategies in this game. Why? Well, if we check the outcomes, we can see that Jim would always prefer to go up if Tim is going left, earning 6, and go down if Tim is going right, earning 4. Tim, on the other hand, would always prefer to go left when Jim is going down, getting 4, and go right when Jim moves up, getting 2.

No pure strategy. Someone will want to change strategies in each scenario. But we worry not as we know now there's a mixed-strategy Nash equilibrium.

Jim's mixed strategy will look like this: the expected utility of Tim of going left and for going right.

$$EU_L = EU_R$$

$$EU_L = f(\sigma_U)$$

$$EU_R = f(\sigma_U)$$

First, we will calculate Tim's left strategy as a function of Jim's strategy sigma up ($f(\sigma_U)$) (See the gray box and bolded, underlined letters.)

Jim↓/Tim →	Left	Right
Up	6; **-2**	-2; 2
Down	-2; **4**	4; -2

Tim will get minus two some percentage of the time (when Jim goes up) and 4 some percentage of the time (when Jim goes down). So, we can write our algorithm as follows:

$$EU_L = \sigma_U(-2) + (1 - \sigma_U)(4)$$

Sigma $_U$ times -2 ($\sigma_U(-2)$ is the probability that Jim will go up and 1 minus sigma $_U$ times 4 $(1 - \sigma_U)(4)$ is the probability of Jim going down. And this collectively is Tim's payoff for choosing to go left.

Jim↓/Tim →	Left	Right
Up	6; -2	-2; **2**
Down	-2; 4	4; **-2**

Let's see the other side and calculate the expected utility of Tim going right. Tim will get 2 some percentage of the time (when Jim goes up) and -2 some percentage of the time (when Jim goes down).

$$EU_R = \sigma_U(2) + (1 - \sigma_U)(-2)$$

Sigma $_U$ times 2 ($\sigma_U(2)$ is the probability that Jim will go up and 1 minus sigma $_U$ times -2 $(1 - \sigma_U)(-2)$ is the probability of Jim going down. And this collectively is Tim's payoff for choosing to go right.

As $EU_L = EU_R$, we can write our equation as follows:

$$\sigma_U(-2) + (1 - \sigma_U)(4) = \sigma_U(2) + (1 - \sigma_U)(-2)$$

Let's solve this equation step-by-step.

Step 1: Simplify both sides of the equation.

$$\sigma_U(-2) + (1 - \sigma_U)(4) = \sigma_U(2) + (1 - \sigma_U)(-2)$$

$$\sigma_U(-2) + (1)(4) + (-\sigma_U)(4) = \sigma_U(2) + (1)(-2) + (-\sigma_U)(-2) \text{ (Distribute)}$$

$$-2\sigma_U + 4 + -4\sigma_U = 2\sigma_U + -2 + 2\sigma_U$$

$$(-2\sigma_U + -4\sigma_U) + (4) = (2\sigma_U + 2\sigma_U) + (-2) \text{ (Combine Like Terms)}$$

$$-6\sigma_U + 4 = 4\sigma_U + (-2)$$

$$-6\sigma_U + 4 = 4\sigma_U + -2$$

Step 2: Subtract $4\sigma_U$ from both sides.

$$-6\sigma_U + 4 - 4\sigma_U = 4\sigma_U + -2 - 4\sigma_U$$

$$-10\sigma_U + 4 = -2$$

Step 3: Subtract 4 from both sides.

$$-10\sigma_U + 4 - 4 = -2 - 4$$

$$-10\sigma_U = -6$$

Step 4: Divide both sides by -10.
$$-10\sigma_U / -10 = -6 / -10$$

$$\sigma_U = 3/5$$

Answer:
$\sigma_U = 3/5$

In other words, when Jim goes up with a probability of 3/5 and down with a probability of 2/5, Tim is indifferent whether to go left or right.

Half of our work is done. Now we need to see if Tim has a mixed strategy that leaves Jim indifferent between up and down.

We ask what the expected utility of up and expected utility of down is as a function of Tim's mixed strategy. Remember, they have to be equal.

$$EU_U = EU_D$$

$$EU_U = f(\sigma_L)$$

$$EU_D = f(\sigma_L)$$

Let's calculate Jim's strategy of going up as a function of Tim's strategy sigma $_L$ (f(σ_L). (See the gray box and bolded, underlined letters.)

Jim↓/Tim →	Left	Right
Up	**6**; -2	**-2**; 2
Down	-2; 4	4; -2

Jim will get 6 some percentage of the time (when Tim goes left) and -2 some percentage of the time (when Tim goes right).

$$EU_U = \sigma_L(6) + (1 - \sigma_L)(-2)$$

Sigma $_L$ times 6 ($\sigma_L(6)$ is the probability that Tim will go left and 1 minus sigma $_L$ times -2 $(1 - \sigma_L)(-2)$ is the probability of Tim going right. And this collectively is Jim's payoff for choosing to go up.

Jim↓/Tim →	Left	Right

Up	6; -2	-2; 2
Down	**-2**; 4	**4**; -2

Let's see the other side and calculate the expected utility of Jim going down. Jim will get -2 some percentage of the time (when Tim goes left) and 4 some percentage of the time (when Tim goes right).

$$EU_D = \sigma_L (-2) + (1 - \sigma_L)(4)$$

Sigma $_L$ times -2 (σ_L (-2) is the probability that Tim will go left and 1 minus sigma $_L$ times 4 $(1 - \sigma_L)(4)$ is the probability of Tim going right. And this collectively is Jim's payoff for choosing to go down.

As $EU_U = EU_D$, we can write our equation as follows:

$$\sigma_L(6) + (1 - \sigma_L)(-2) = \sigma_L(-2) + (1 - \sigma_L)(4)$$

Let's solve this equation step-by-step.

Step 1: Simplify both sides of the equation.

$$\sigma_L(6) + (1-\sigma_L)(-2) = \sigma_L(-2) + (1-\sigma_L)(4)$$

$$\sigma_L(6) + (1)(-2) + (-\sigma_L)(-2) = \sigma_L(-2) + (1)(4) + (-\sigma_L)(4) \text{ (Distribute)}$$

$$6\sigma_L + -2 + 2\sigma_L = -2\sigma_L + 4 + -4\sigma_L$$

$$(6\sigma_L + 2\sigma_L) + (-2) = (-2\sigma_L + -4\sigma_L) + (4) \text{ (Combine Like Terms)}$$

$$8\sigma_L + (-2) = -6\sigma_L + 4$$

$8\sigma_L + -2 = -6\sigma_L + 4$

Step 2: Add $6\sigma_L$ to both sides.

$8\sigma_L + -2 + 6\sigma_L = -6\sigma_L + 4 + 6\sigma_L$

$14\sigma_L + -2 = 4$

Step 3: Add 2 to both sides.

$14\sigma_L + -2 + 2 = 4 + 2$

$14\sigma_L = 6$

Step 4: Divide both sides by 14.

$14\sigma_L / 14 = 6 / 14$

$\sigma_L = 3/7$

Answer:

$\sigma_L = 3/7$

In other words, when Tim goes left with a probability of 3/7 and right with a probability of 4/7, Jim is indifferent whether to go up or down.

Let's put all of our information into our game matrix:

Jim↓/ Tim →	**Left 3/7**	**Right 4/7**
Up 3/5	6; -2	-2; 2
Center	0; 0	0; 0
Down 2/5	-2; 4	4; -2

The strictly dominant mixed-strategy Nash equilibrium is: Jim plays up with probability 3/5 and down 2/5. Tim goes left with a probability of 3/7 and right 4/7. We already established that there is no pure-strategy Nash equilibrium for Jim or Tim. We also know that center is a strictly dominated strategy.

Conclusively, this strictly dominant mixed-strategy Nash equilibrium is much better than going for center.

Weak Dominance

What is weak dominance and why do I want to talk about it? I bet weak dominance doesn't sound like an appealing strategy. It is and it isn't. At the end of the day, it is still a form of dominance—it's better to play this out than a strictly dominated strategy. And, in some cases, it can yield better payoffs for one of the players.

Recall the practice of iterated elimination of strictly dominated strategies. We want to make sure to get rid of those strategies as playing them would be irrational. And we established that in simultaneous games (the types of games we examined so far), players are rational and well-informed decision-makers. They are always trying to find a strategy in Nash equilibrium—as this means they found a strategy they have no incentive to deviate from.

Jim↓/Tim →	**Left**	**Right**
Up	4; **2**	1; **1**
Down	3; 3	3; 3

This is a simple game where Jim has two strategies, up and down, and Tim has two strategies, left and right. Tim knows that if Jim goes up, Tim is better off if he chooses left—2 is bigger than 1. (See the bolded, underlined numbers in the matrix above.)

Jim↓/Tim →	**Left**	**Right**
Up	4; 2	1; 1
Down	3; **3**	3; **3**

However, if Jim goes down, Tim is indifferent between going left or right because his payoff is 3 in either case. (See the bolded, underlined numbers in the matrix above.)

What does this information tell us? That, for Tim, left sometimes is just as good of an option as right and sometimes left is a better option than right. In other words, *left weakly dominates right*. If left strictly dominated right, going left for Tim always had to be better. But with the equal payoff of Jim going down makes left just weakly dominant.

If we wanted to find a Nash equilibrium in the weakly dominant segment, we could discard right—as we established, it is *weakly dominated.*

Jim ↓ / Tim →	**Left**
Up	4; 2
Down	3; 3

We would have this matrix. Here, Jim would like to maximize his benefit, so he would go up (because 4 is bigger than 3). We could then establish that the weakly dominant Nash equilibrium is up-left.

But is this the only Nash equilibrium of the game? I'm sure you're thinking "No ...?" because of the way I posed the question. You are right! There is another Nash equilibrium—and this is the beauty of weakly dominant strategies.

Jim↓/Tim →	**Left**	**Right**
Up	4; 2	1; 1
Down	3; 3	**3; 3**

To find our other Nash equilibrium, we have to look at the original game. The risk of discarding weakly dominated strategies is that we might eliminate Nash equilibria. Take a look at the gray box and the bolded, highlighted numbers.

Can either Jim or Tim deviate from that outcome and get better results? No, they cannot. If Jim changes strategies and goes up, he actually is worse off, as 1 is less than 3. Can Tim profitably deviate from his outcome? No, he cannot. His payoff won't change if he goes left. Therefore right-down is also a Nash equilibrium.

Moreover, it is in Tim's best interest to pursue this Nash equilibrium when he can. His payoff in playing right when Jim plays down is higher than to play left when Jim plays up, as 3 is bigger than 2. The weakly dominated Nash equilibrium in Tim's case is a more profitable strategy. It makes sense for him to commit to it and try to pull Jim into playing down.

Weak Dominance in the Real World

The Battle of the Bismarck Sea happened in 1943 in the South West Pacific Area during World War II. The US Fifth Air Force, strengthened by the Royal Australian Air Force, attacked a convoy carrying Japanese troops in Lae, New Guinea. The attack resulted in the annihilation of the Japanese forces in the region.

Earlier, in 1942, the Japanese Imperial General Headquarters decided to station the convoy in New Guinea to strengthen their position in the South West Pacific. They wanted to move some 6,900 troops from Rabaul directly to Lae. The Japanese knew that their plan involved a high risk. The Allied air forces in the region was strong. Yet moving the forces seemed to be worth the risk at the time, otherwise the troops had to be deployed further away, needing to march through the hazardous terrain of New Guinea.[lvi]

Why am I telling this story? Because it is a great example of weak dominance and the elimination of dominated strategies. In this game, as you can see on the matrix below, Lieutenant General George Kenney had no dominant strategy.

Kenney ↓ / Japanese →	**North**	**South**
North	2; -2	2; -2
South	1; -1	3; -3

The Japanese are indifferent between north and south if Kenney moves north. But if Kenney moves south, the Japanese would want to go north instead of south as -1 is better than -2. In other words, north is a weakly dominant strategy from the Japanese perspective.

Now we need to eliminate dominated strategies. We established that, for the Japanese strategy, north weakly dominates south. Thus, we remove the south strategy for the Japanese.

Kenney↓/Japanese →	**North**
North	2; -2
South	1; -1

Now that we know the Japanese will be going north, we can see what Kenney's best response is. The answer is that the north strategy strictly dominates the south strategy, because 2 is bigger than 1. We can then eliminate south from Kenney's strategy and conclude that north-north is the weakly dominant Nash equilibrium.[lvii]

Kenney↓/Japanese →	**North**
North	2; -2

Chapter 7: Curious Tales from the Land of Game Theory

Infinitely Many Equilibria[lviii]

Most games we have played thus far had either pure- or mixed-strategy Nash equilibria. But what if I told you there can be infinitely many equilibria in a game? Sounds crazy, right? Well, even if it does, it's possible. Let's see how analyzing the following game where Jim and Tim both have two strategies. Jim can go either up or down and Tim can go either left or right.

Jim↓/Tim →	Left	Right

Up	10; 1	5; 2
Down	3; -4	5; -3

If we start fishing for strictly dominant strategies, we will find that right strictly dominates left from Tim's perspective. If Jim decides to go up, Tim will want to go right as 2 is bigger than 1. If Jim goes down, Tim still will want to go right as -3 is better than -4. Using the iterated elimination of strictly dominated strategies, we can remove the left side.

Jim ↓ / Tim →	**Right**
Up	5; 2
Down	5; -3

We can see that if Tim plays right, Jim is indifferent between going up or down.

He will get 5 either way. Let's try to find some Nash equilibria.

Jim ↓ / Tim →	**Right**
Up	5; 2

Is this a Nash equilibrium? Try to answer this question before you read further.

Did you answer?

What's the answer?

Okay. I gave you some space to think. The answer is yes. This is a Nash equilibrium; a pure-strategy Nash equilibrium, more precisely. Why? Because neither player can deviate profitably. Tim would get 1 instead of 2 if he went left. Jim would get the same 5 if he went down.

Neither player can do better by changing strategies. We found a pure-strategy Nash equilibrium. But is this the only one?

Again, before I answer, go back to the main game matrix and try to find another Nash equilibrium.

Were you successful?

Okay, I'm going to tell you now. The last chance to find it yourself.

Jim ↓ / Tim →	**Right**
Down	5; -3

This is also a pure-strategy Nash equilibrium. Why? Because neither player can benefit from changing strategies. Tim would get -4 if he went left. Jim would get the same 5 if he went up. It turns out that the strictly dominant strategy's both options are in fact pure-strategy Nash equilibria.

Big "but" coming ... But let's not forget about possible mixed strategies in this game!

In this game, we have a unique type of mixed strategy called partially mixed strategy.

Jim ↓ / Tim →	**Right**
Up (p)	5; 2
Down (1-p)	5; -3

In a partially mixed-strategy Nash equilibrium, one player plays a pure strategy and the other one a mixed strategy. In our game, Tim is the one who is playing the pure strategy: He's always going right. But Jim follows a mixed strategy where he chooses up with a probability p and down with a probability 1-p. Any number of p can make this into an equilibrium. Why?

Tim doesn't want to change his strategy. And because Jim is indifferent, he's always getting 5 no matter what, meaning he can't deviate and get a better payoff. No matter what he does, he will gain 5, period.

Because of these two factors, Jim can mix how many times he goes up or down with no risk. He can go up 1 percent of the time and down 99 percent of the time. Or he can go up 2 percent of the time and down 98 percent of the time. You get my point. As there are infinitely many numbers between 0 and 100, he can have infinitely many options for a mixed strategy and therefore infinitely many equilibria.

Isn't this *odd*? Well, this question is a Freudian slip about what I will talk about next.

Even Number of Equilibria

I don't know if you paid attention, but all the games we've explored so far had an odd number of equilibria. The prisoner's dilemma had one pure-strategy Nash equilibrium. The matching pennies had one, and the mixed-strategy algorithm game had one mixed-strategy Nash equilibrium. The stag hunt and the battle of the sexes had three: two pure and one mixed.

Games with even number of equilibria or with an infinite number of equilibria are actually very rare. And when they happen, that's usually because of the weak dominance we learned about in the previous chapter.

The last game we will explore in this book is a game with an even number of equilibria called the free money game.

In this game, Jim and his brother, Tim, are offered some money. The only thing the brothers have to do is unanimously, simultaneously, and blindly approve receiving the free money. If Jim and Tim both say yes to getting the money, they both will. But if one of them says no, neither of them gets any money.

I know, who on earth would say no to free money? But let's assume that in some cases a shady figure offers them the money and a conscience wakes up in one or both of the brothers. "How did this guy get the money he's giving us? What if there is blood attached to it? I don't want to be part of something illegal. I will say no."

Jim↓/ Tim →	Yes	No
Yes	1; 1	0; 0
No	0; 0	0; 0

This is how the free money game looks like. As you can see, if Jim and Tim both say yes, they get one unit of, say, one thousand dollars. If either or both say no, they get nothing. Yes-yes, therefore, is a pure-strategy Nash equilibrium. The brothers can't change their strategy and get a better outcome.

Jim ↓ / Tim →	**No**
No	0; 0

But no-no is also a pure-strategy Nash equilibrium because they can't profitably deviate from their response. If Jim said yes, they would still get nothing. The same is true if Tim says yes. Neither of them gets anything.

So far, we have two pure-strategy Nash equilibria. But we need to make sure that there is no mixed-strategy Nash equilibrium in this game to be able to call this game one with an even number of equilibria.

Jim↓/ Tim →	Yes	No
Yes	1; 1	0; 0
No	0; 0	0; 0

To discover whether there is a mixed-strategy Nash equilibrium, let's assume that Jim is mixing his strategy. Namely, sometimes he says yes and sometimes he says no. Tim is creating his best response to Jim's mixed strategy. Jim has to stick to yes, no matter what. Because if he says no, he will get 0 for sure. His best chance to get free money is to say yes regardless of what Jim does. In the instances when Jim says no as a part of his mixed strategy, Tim will get 0 as well, even if he said yes. But whenever Jim says yes, Tim will get 1. So, Tim has to play a pure strategy (say yes) in response to Jim's mixed strategy.

But if Tim plays yes as a pure strategy, Jim will not want to play a mixed strategy because, then, whenever he says no, he robs himself of free money. It's not a profitable strategy to ever play no as long as Tim purely plays yes. This is why there is no mixed-strategy Nash equilibrium in this game. Neither of them can mix and deviate profitably.

Yes, technically, is a weakly dominant strategy in this game. In some cases, it pays better than no. (When both players say yes.) And in other cases, it brings the same results as no. (When one player says no.)

The bottom line in this chapter is that even or infinite number of equilibria are rare.

Whenever you find a game with an even number of equilibria, make sure to check for mixed strategies with the mixed-strategy algorithm. Or if you did, recalculate everything you did, just to make sure all is correct. Look for weakly dominated strategies, as well. If you find one, chances are high that you are correct and your game has an even or infinite number of Nash equilibria.

The free money game sounded a bit silly, right? But you would be shocked how often this scenario can play out in the real world. Suppose a country with a stronger military power tries to force trade deals with a weaker country.

Let's assume there is 10 billion dollars at stake. The stronger country is in a power position, so it "suggests" to the weaker country that the distribution of this money should be 9 billion vs. 1 billion. Take it or leave it! Since the entire deal depends on the natural resources of the weaker country, they will only profit if both of them say yes. We are talking about a similar game as before, only now it looks like this:

Strong ↓ / Weak →	**Yes**	**No**
Yes	9; 1	0; 0
No	0; 0	0; 0

The weak country can protest as much as it wants about the distribution not being fair, about getting the shorter end of the stick, about how this deal is just an exploitation project—if they say no, they fall 1 billion dollars short. We assume here that the weak country doesn't have the infrastructure to significantly profit from its natural resources. They need the stronger country. The stronger country also needs the weak country for their resources. But knowing game theory and being in a power position, they feel comfortable offering a ridiculous deal such as the one above. They know that the weak country is a rational player, so they will say yes. Begrudgingly, but they will say yes.

The strong country has other weapons in their arsenal, just in case the weak country says no. It can choose other type of persuasion strategies, such as invasion. But that's another game. Here, I only wanted to illustrate how important and useful this game can be in real-world negotiations.

This game, coined by Robert Yisrael Aumann as the Blackmailer's Paradox,[lix] is actually a game in extensive form. I just transformed it into a normal-form simultaneous game here to give you an idea of the real-world utility of such deals. Also, to spark your curiosity for the sequel of this book, where we will delve deep into extensive-form games.

Exercise

I don't know if you noticed, but I said that the stag hunt game had three Nash equilibria. But if you recall, we only found two in this book. Your first exercise will be to find the third mixed-strategy Nash equilibrium.

Jim↓/Tim →	Stag	Hare
Stag	4; 4	0; 2
Hare	2; 0	1; 1

Let's recall our stag hunt game. We concluded that stag-stag and hare-hare are both pure-strategy Nash equilibria. But I spoiled for you that there is a mixed-strategy Nash equilibrium.

Your task is to use the mixed-strategy algorithm to find it. Let me help you a little bit by giving a name for each strategy so you can more easily set up the expected utility equations. Here you go:

Jim↓/Tim →	**Stag (TS)**	**Hare (TH)**
Stag (SJ)	4; 4	0; 2
Hare (SH)	2; 0	1; 1

$$EU_{TS} = EU_{TH}$$

$$EU_{TS} = f(\sigma_{SJ})$$

$$EU_{TH} = f(\sigma_{SJ})$$

Good luck!

(If you want the solution for this game, shoot me an email: albertrutherfordbooks@gmail.com)

Conclusion

While game theory models are interesting and develop the strategic mind, it's important to remember the nature of models in general: They are a simplification of the real world. This stands true in the case of game theory models, too. Real-world issues are usually much more complicated than a model can capture. No mathematical algorithm can capture their full complexity.

Even so, being able to think ahead of our competitors, conclude successful negotiations, and knowing how to secure the best possible outcome for ourselves gives us an advantage in our lives. Negotiations, agreements, and competitive-advantage seeking happens on a large scale—such as between countries—or on a small scale—such as between us and our spouse, boss, or business partners. Being skilled in the art and science of strategy gives us a competitive advantage as decision-makers on any level.

This book has been a primer to game theory; an introductory study to get your strategic juices flowing. Remember, every piece of advice and practice in a book is worth only as much as you make of it. So, if you didn't solve the exercise in the previous chapter, go back and do it now.

Best of luck, my strategist friend!

A.R.

For your FREE GIFT: The Art of Asking Powerful Questions in the World of Systems visit www.albertrutherford.com.

Resources

Aumann, Robert J. (1976). "Agreeing to Disagree". The Annals of Statistics. Institute of Mathematical Statistics. 4 (6): 1236–1239. doi:10.1214/aos/1176343654. ISSN 0090-5364. JSTOR 2958591.

Aumann, Yisrael. (n.d.a). The Blackmailer's Paradox. Temple Beth Sholom. Retrieved March 24, 2021, from https://test.tbshamden.com/odds-a-ends/the-blackmailers-paradox/

Cazals, C. (2016, November 30). *How game theory affects your everyday life*. The London Globalist. https://thelondonglobalist.org/how-game-theory-affects-your-everyday-life/

Chandrasekaran, R. (n.d.). *Cooperative Game Theory*. The University of Texas. Retrieved March 25, 2021, https://personal.utdallas.edu/~chandra/documents/6311/coopgames.pdf

Chen, J. (2021, March 3). *Nash Equilibrium*. Investopedia.

https://www.investopedia.com/terms/n/nash-equilibrium.asp

Course Hero. (n.d.). *Chapter 14 Summary and Review - Monopolistic Competition.* Retrieved March 24, 2021, from https://www.coursehero.com/file/p460btv/Nash-Equilibrium-is-a-set-of-strategies-or-actions-in-which-each-firm-does-the/

Crossman, A. (2019, March 1). *What is Game Theory?: An Overview of the Sociological Concept.* ThoughtCo. https://www.thoughtco.com/game-theory-3026626

Dieudonné, J. (2008). "Von Neumann, Johann (or John)". In Gillispie, C. C. (ed.). Complete Dictionary of Scientific Biography. 14 (7th ed.). Detroit: Charles Scribner's Sons. pp. 88–92 Gale Virtual Reference Library. ISBN 978-0-684-31559-1. OCLC 187313311

Fang, C., Kimbrough, S. O., Pace, S., Valluri, A., & Zheng, Z. (2002). On Adaptive Emergence of Trust Behavior in the Game of Stag Hunt. *Group Decision and Negotiation, 11*(6), 449–467. https://doi.org/10.1023/A:1020639132471

Game theory. (2021, March 13). In *Wikipedia*. https://en.wikipedia.org/w/index.php?title=Game_theory&oldid=1011854501

Ganti, A. (2021, February 22). *Rational Choice Theory*. Investopedia. https://www.investopedia.com/terms/r/rational-choice-theory.asp

Harsanyi, John C. (1961). "On the Rationality Postulates underlying the Theory of Cooperative Games". The Journal of Conflict Resolution. 5 (2): 179–196. doi:10.1177/002200276100500205. S2CID 220642229.

Heilbroner, R. L. (n.d.). Adam Smith. In *Encyclopedia Britannica*. Retrieved March 24, 2021, from https://www.britannica.com/biography/Adam-Smith

Kenton, W. (2020, September 28). *Behavioral Economics*. Investopedia. https://www.investopedia.com/terms/b/behavioraleconomics.asp

Kolbert, E. (2008, February 17). *What Was I Thinking?*. The New Yorker. https://www.newyorker.com/magazine/2008/02/25/what-was-i-thinking

Mirriam-Webster. (n.d.). Game. In *Mirriam-Webster.com dictionary*. Retrieved

March 24, 2021, from https://www.merriam-webster.com/dictionary/game

Mirriam-Webster. (n.d.). Theory. In *Mirriam-Webster.com dictionary.* Retrieved March 24, 2021, from https://www.merriam-webster.com/dictionary/theory

Murray, Williamson; Millett, Allan R. (2001). A War To Be Won: Fighting the Second World War. Cambridge, Massachusetts: Belknap Press. ISBN 0-674-00680-1.

Mycielski, Jan (1992). "Games with Perfect Information". Handbook of Game Theory with Economic Applications. 1. pp. 41–70. doi:10.1016/S1574-0005(05)80006-2. ISBN 978-0-4448-8098-7.

Nasar, Sylvia (1998) A Beautiful Mind, Simon & Schuster. ISBN 0-684-81906-6.

Osborne, Martin J.; Rubinstein, Ariel (12 Jul 1994). A Course in Game Theory. Cambridge, MA: MIT. p. 14. ISBN 9780262150415.

Oppenheimer, D.M., & Monin, B. (2009). The retrospective gambler's fallacy: Unlikely events, constructing the past, and multiple universes. *Judgment and*

Decision Making, vol. 4, no. 5, pp. 326-334

Owen, Guillermo (1995). Game Theory: Third Edition. Bingley: Emerald Group Publishing. p. 11. ISBN 978-0-12-531151-9.

Picardo, E. (2019, May 19). *How Game Theory Strategy Improves Decision Making.* Investopedia. https://www.investopedia.com/articles/investing/111113/advanced-game-theory-strategies-decisionmaking.asp

Picardo, E. (2020, January 22). *The Prisoner's Dilemma in Business and the Economy.* Investopedia. https://www.investopedia.com/articles/investing/110513/utilizing-prisoners-dilemma-business-and-economy.asp

Policonomics. (n.d.a). *Game theory II: Dominant strategies.* Retrieved March 24, 2021, from https://policonomics.com/lp-game-theory2-dominant-strategy/

Policonomics. (n.d.b). *Game theory II: Simultaneous games.* Retrieved March 24, 2021, from https://policonomics.com/lp-game-theory2-simultaneous-game/

Poundstone, William (1993). Prisoner's Dilemma (1st Anchor Books ed.). New York: Anchor. ISBN 0-385-41580-X.

Prisner, E. (2014). *Game Theory through Examples* [eBook edition]. Classroom Resource Materials. https://www.maa.org/sites/default/files/pdf/ebooks/GTE_sample.pdf.

Rasmusen, Eric (2007). Games and Information (4th ed.). ISBN 978-1-4051-3666-2.

Rubinstein, A. (1991). "Comments on the interpretation of Game Theory". Econometrica. 59 (4): 909–924. doi:10.2307/2938166. JSTOR 2938166.

ScienceDirect. (2014). Ultimatum Game - an overview. Retrieved March 24, 2021, from https://www.sciencedirect.com/topics/neuroscience/ultimatum-game

Shoham, Yoav; Leyton-Brown, Kevin (15 December 2008). *Multiagent Systems: Algorithmic, Game-Theoretic, and Logical Foundations*. Cambridge University Press. ISBN 978-1-139-47524-2.

Shor, M. (2005a, August 12). *Deadlock*. Gametheory.net.

https://www.gametheory.net/dictionary/Games/Deadlock.html

Shor, M. (2005b, August 12). *Symmetric Game.* Gametheory.net. https://www.gametheory.net/dictionary/Games/SymmetricGame.html

Shor, M. (2005c, August 15). *Non-Cooperative Game.* Gametheory.net. https://www.gametheory.net/dictionary/Non-CooperativeGame.html

Shor, M. (2005d, August 15). *Pareto Optimal.* Gametheory.net. https://www.gametheory.net/dictionary/ParetoOptimal.html

Singh, S. (1998, June 14). Between Genius and Madness. *The New York Times.* https://archive.nytimes.com/www.nytimes.com/books/98/06/14/reviews/980614.14singht.html

Spaniel, W. (2011). *Game Theory 101: The Complete Textbook.* Self-published.

Spaniel, W. (2020, June 24). *Best Responses and Safety in Numbers.* Game Theory 101. http://gametheory101.com/courses/game-theory-101/best-responses/

SportMob. (2020, October 17). *Worst Penalty Misses in football History.*

https://sportmob.com/en/article/897358-worst-penalty-misses-in-football-history

Yu, W. (n.d.). *De Beers – Rulers of the Diamond Industry: The Rise and Fall of a Monopoly.* Retrieved March 24, 2021, from https://are.berkeley.edu/~sberto/DeBeersDiamondIndustry.pdf

Williams, Paul D. (2013). Security Studies: an Introduction (second ed.). Abingdon: Routledge. pp. 55–56.

Endnotes

[i] Merriam-Webster. (n.d.). Game. In Merriam-Webster.com dictionary. Retrieved March 24, 2021, from https://www.merriam-webster.com/dictionary/game

[ii] Merriam-Webster. (n.d.). Theory. In Merriam-Webster.com dictionary. Retrieved March 24, 2021, from https://www.merriam-webster.com/dictionary/theory

[iii] Myerson, Roger B. (1991). Game Theory: Analysis of Conflict, Harvard University Press, p. 1. Chapter-preview links, pp. vii–xi.

[iv] Aumann, Robert J. (1976). "Agreeing to Disagree". The Annals of Statistics. Institute of Mathematical Statistics. 4 (6): 1236–1239. doi:10.1214/aos/1176343654. ISSN 0090-5364. JSTOR 2958591.

[v] Picardo, E. (2019, May 19). How Game Theory Strategy Improves Decision Making. Investopedia. https://www.investopedia.com/articles/investing/111113/advanced-game-theory-strategies-decisionmaking.asp

[vi] Kolbert, E. (2008, February 17). What Was I Thinking?. The New Yorker. https://www.newyorker.com/magazine/2008/02/25/what-was-i-thinking

[vii] Ganti, A. (2021, February 22). Rational Choice Theory. Investopedia. https://www.investopedia.com/terms/r/rational-choice-theory.asp

[viii] Kenton, W. (2020, September 28). Behavioral Economics. Investopedia. https://www.investopedia.com/terms/b/behavioraleconomics.asp

[ix] Dieudonné, J. (2008). "Von Neumann, Johann (or John)". In Gillispie, C. C. (ed.). Complete Dictionary of Scientific Biography. 14 (7th ed.). Detroit: Charles Scribner's Sons. pp. 88–92 Gale Virtual Reference Library. ISBN 978-0-684-31559-1. OCLC 187313311

[x] Schotter, Andrew (1992). "Oskar Morgenstern's Contribution to the Development of the Theory of Games". In Weintraub, E. Roy (ed.). Toward a History of Game Theory. Durham: Duke University Press. pp. 95–112. ISBN 0-8223-1253-0.

[xi] Chandrasekaran, R. (n.d.). Cooperative Game Theory. The University of Texas. Retrieved March 25, 2021, https://personal.utdallas.edu/~chandra/documents/6311/coopgames.pdf

[xii] Shor, M. (2005c, August 15). Non-Cooperative Game. Gametheory.net.

https://www.gametheory.net/dictionary/Non-CooperativeGame.html

[xiii] Shor, M. (2005b, August 12). Symmetric Game. Gametheory.net. https://www.gametheory.net/dictionary/Games/SymmetricGame.html

[xiv] Harsanyi, John C. (1961). "On the Rationality Postulates underlying the Theory of Cooperative Games". The Journal of Conflict Resolution. 5 (2): 179–196. doi:10.1177/002200276100500205. S2CID 220642229.

[xv] ScienceDirect. (2014). Ultimatum Game - an overview. Retrieved March 24, 2021, from https://www.sciencedirect.com/topics/neuroscience/ultimatum-game

[xvi] Owen, Guillermo (1995). Game Theory: Third Edition. Bingley: Emerald Group Publishing. p. 11. ISBN 978-0-12-531151-9.

[xvii] Mycielski, Jan (1992). "Games with Perfect Information". Handbook of Game Theory with Economic Applications. 1. pp. 41–70. doi:10.1016/S1574-0005(05)80006-2. ISBN 978-0-4448-8098-7.

[xviii] Prisner, E. (2014). Game Theory through Examples [eBook edition]. Classroom Resource Materials. https://www.maa.org/sites/default/files/pdf/ebooks/GTE_sample.pdf.

[xix] Rasmusen, Eric (2007). Games and Information (4th ed.). ISBN 978-1-4051-3666-2.

[xx] Shoham, Yoav; Leyton-Brown, Kevin (15 December 2008). *Multiagent Systems: Algorithmic, Game-Theoretic, and Logical Foundations*. Cambridge University Press. ISBN 978-1-139-47524-2.

[xxi] Williams, Paul D. (2013). Security Studies: an Introduction (second ed.). Abingdon: Routledge. pp. 55–56.

[xxii] Game theory. (2021, March 13). In *Wikipedia*. https://en.wikipedia.org/wiki/Game_theory#/media/File:Ultimatum_Game_Extensive_Form.svg

[xxiii] Yu, W. (n.d.). De Beers – Rulers of the Diamond Industry: The Rise and Fall of a Monopoly. Retrieved March 24, 2021, from https://are.berkeley.edu/~sberto/DeBeersDiamondIndustry.pdf

[xxiv] Poundstone, William (1993). Prisoner's Dilemma (1st Anchor Books ed.). New York: Anchor. ISBN 0-385-41580-X.

[xxv] An Overview of Game Theory in Sociology - ThoughtCo. https://www.thoughtco.com/game-theory-3026626

[xxvi] Picardo, E. (2020, January 22). The Prisoner's Dilemma in Business and the Economy. Investopedia. https://www.investopedia.com/articles/investin

g/110513/utilizing-prisoners-dilemma-business-and-economy.asp

[xxvii] Shor, M. (2005a, August 12). Deadlock. Gametheory.net. https://www.gametheory.net/dictionary/Games/Deadlock.html

[xxviii] Shor, M. (2005d, August 15). Pareto Optimal. Gametheory.net. https://www.gametheory.net/dictionary/Pareto Optimal.htmll

[xxix] Spaniel, W. (2011). Game Theory 101: The Complete Textbook. Self-published.

[xxx] Singh, S. (1998, June 14). Between Genius and Madness. The New York Times. https://archive.nytimes.com/www.nytimes.com/books/98/06/14/reviews/980614.14singht.html

[xxxi] Nasar, Sylvia (1998) A Beautiful Mind, Simon & Schuster. ISBN 0-684-81906-6.

[xxxii] Singh, S. (1998, June 14). Between Genius and Madness. The New York Times. https://archive.nytimes.com/www.nytimes.com/books/98/06/14/reviews/980614.14singht.html

[xxxiii] Nasar, Sylvia (1998) A Beautiful Mind, Simon & Schuster. ISBN 0-684-81906-6.

[xxxiv] Course Hero. (n.d.). *Chapter 14 Summary and Review - Monopolistic Competition.* Retrieved March 24, 2021, from https://www.coursehero.com/file/p460btv/Nash-Equilibrium-is-a-set-of-strategies-or-actions-in-which-each-firm-does-the/

[xxxv] Cazals, C. (2016, November 30). How game theory affects your everyday life. The London Globalist. https://thelondonglobalist.org/how-game-theory-affects-your-everyday-life/

[xxxvi] Heilbroner, R. L. (n.d.). Adam Smith. In Encyclopedia Britannica. Retrieved March 24, 2021, from https://www.britannica.com/biography/Adam-Smith

[xxxvii] Cazals, C. (2016, November 30). How game theory affects your everyday life. The London Globalist. https://thelondonglobalist.org/how-game-theory-affects-your-everyday-life/

[xxxviii] Spaniel, W. (2011). Game Theory 101: The Complete Textbook. Self-published.

[xxxix] Fang, C., Kimbrough, S. O., Pace, S., Valluri, A., & Zheng, Z. (2002). On Adaptive Emergence of Trust Behavior in the Game of Stag Hunt. Group Decision and Negotiation, 11(6), 449–467. https://doi.org/10.1023/A:1020639132471

[xl] Spaniel, W. (2011). Game Theory 101: The Complete Textbook. Self-published.

[xli] Spaniel, W. (2011). Game Theory 101: The Complete Textbook. Self-published.

[xlii] Spaniel, W. (2020, June 24). Best Responses and Safety in Numbers. Game Theory 101. http://gametheory101.com/courses/game-theory-101/best-responses//

[xliii] Chen, J. (2021, March 3). Nash Equilibrium. Investopedia. https://www.investopedia.com/terms/n/nash-equilibrium.asp

[xliv] Osborne, Martin J.; Rubinstein, Ariel (12 Jul 1994). A Course in Game Theory. Cambridge, MA: MIT. p. 14. ISBN 9780262150415.

[xlv] Spaniel, W. (2011). Game Theory 101: The Complete Textbook. Self-published.

[xlvi] Spaniel, W. (2011). Game Theory 101: The Complete Textbook. Self-published.

[xlvii] Oppenheimer, D.M., & Monin, B. (2009). The retrospective gambler's fallacy: Unlikely events, constructing the past, and multiple universes. *Judgment and Decision Making, vol. 4, no. 5,* pp. 326-334

[xlviii] Spaniel, W. (2011). Game Theory 101: The Complete Textbook. Self-published.

[xlix] Spaniel, W. (2011). Game Theory 101: The Complete Textbook. Self-published.

[l] SportMob. (2020, October 17). Worst Penalty Misses in football History. https://sportmob.com/en/article/897358-worst-penalty-misses-in-football-history

[li] Rubinstein, A. (1991). "Comments on the interpretation of Game Theory". Econometrica. 59 (4): 909–924. doi:10.2307/2938166. JSTOR 2938166.

[lii] Policonomics. (n.d.b). Game theory II: Simultaneous games. Retrieved March 24,

2021, from https://policonomics.com/lp-game-theory2-simultaneous-game/

[liii] Spaniel, W. (2011). Game Theory 101: The Complete Textbook. Self-published.

[liv] Spaniel, W. (2011). Game Theory 101: The Complete Textbook. Self-published.

[lv] Spaniel, W. (2011). Game Theory 101: The Complete Textbook. Self-published.

[lvi] Murray, Williamson; Millett, Allan R. (2001). A War To Be Won: Fighting the Second World War. Cambridge, Massachusetts: Belknap Press. ISBN 0-674-00680-1.

[lvii] Policonomics. (n.d.a). Game theory II: Dominant strategies. Retrieved March 24, 2021, from https://policonomics.com/lp-game-theory2-dominant-strategy/

[lviii] Spaniel, W. (2011). Game Theory 101: The Complete Textbook. Self-published.

[lix] Aumann, Yisrael. (n.d.a). The Blackmailer's Paradox. Temple Beth Sholom. Retrieved March 24, 2021, from https://test.tbshamden.com/odds-a-ends/the-blackmailers-paradox/

Printed in Poland
by Amazon Fulfillment
Poland Sp. z o.o., Wrocław